SpringerBriefs in Electrical and Computer Engineering

Computational Electromagnetics

Series editor

Rakesh Mohan Jha, Bangalore, India

More information about this series at http://www.springer.com/series/13885

Hema Singh · R. Chandini
Rakesh Mohan Jha

Low Profile Conformal Antenna Arrays on High Impedance Substrate

 Springer

Hema Singh
Centre for Electromagnetics
CSIR-National Aerospace Laboratories
Bangalore, Karnataka
India

Rakesh Mohan Jha
Centre for Electromagnetics
CSIR-National Aerospace Laboratories
Bangalore, Karnataka
India

R. Chandini
Centre for Electromagnetics
CSIR-National Aerospace Laboratories
Bangalore, Karnataka
India

ISSN 2191-8112   ISSN 2191-8120   (electronic)
SpringerBriefs in Electrical and Computer Engineering
ISSN 2365-6239   ISSN 2365-6247   (electronic)
SpringerBriefs in Computational Electromagnetics
ISBN 978-981-287-762-8   ISBN 978-981-287-763-5   (eBook)
DOI 10.1007/978-981-287-763-5

Library of Congress Control Number: 2015947941

Springer Singapore Heidelberg New York Dordrecht London

Printed on acid-free paper

Springer Science+Business Media Singapore Pte Ltd. is part of Springer Science+Business Media
(www.springer.com)

To Dr. Sudhakar K. Rao

*In Memory of Dr. Rakesh Mohan Jha
Great scientist, mentor, and excellent
human being*

Dr. Rakesh Mohan Jha was a brilliant contributor to science, a wonderful human being, and a great mentor and friend to all of us associated with this book. With a heavy heart we mourn his sudden and untimely demise and dedicate this book to his memory.

Preface

The High Impedance Surface (HIS) is a preferred substrate for low profile antenna design, owing to its unique boundary conditions. Such substrates permit radiating elements to be printed on them, without any disturbance to the radiation characteristics. Moreover, HIS provides improved impedance matching, enhanced bandwidth, increased broadside directivity owing to total reflection from the reactive surface, and high input impedance. This book presents EM design and analysis of dipole antenna array over high impedance substrate. Different configurations of HIS such as array of metallic dogbones or square patches are considered for the array design on planar and nonplanar high-impedance surface. The HIS unit cell design consists of single and double layer of conductors over a PEC ground plane-backed dielectric substrate. The antenna elements are placed on a top surface of metamaterial/HIS consisting of tightly coupled metallic dogbones. The substrate is backed by a perfectly conducting ground plane. The antenna design consisting of a multi-layered substrate and PEC ground plane is limited within sub-wavelengths. This book presents the EM design and analysis of cylindrical dipole, printed dipole, and folded dipole over single and double layered square-patch based-HIS and dogbone-based HIS. The performance of array is analyzed in terms of return loss and radiation pattern. The overall design performance depends on both the radiating element and the HIS parameters. The array design is extended for nonplanar cylindrical HIS. This book helps reader to efficiently design arbitrary low profile antenna array over planar and nonplanar HIS based substrate.

Hema Singh
R. Chandini
Rakesh Mohan Jha

Acknowledgments

We would like to thank Mr. Shyam Chetty, Director, CSIR-National Aerospace Laboratories, Bangalore for his permissions and support to write this SpringerBrief.

We would also like to acknowledge valuable suggestions from our colleagues at the Centre for Electromagnetics, Dr. R.U. Nair, Dr. Shiv Narayan, Dr. Balamati Choudhury, and Mr. K.S. Venu during the course of writing this book. We express our sincere thanks to Mr. Harish S. Rawat, Ms. Neethu P.S., Mr. Umesh V. Sharma, and Mr. Bala Ankaiah, the project staff at the Centre for Electromagnetics, for their consistent support during the preparation of this book.

But for the concerted support and encouragement from Springer, especially the efforts of Suvira Srivastav, Associate Director, and Swati Meherishi, Senior Editor, Applied Sciences & Engineering, it would not have been possible to bring out this book within such a short span of time. We very much appreciate the continued support by Ms. Kamiya Khatter and Ms. Aparajita Singh of Springer towards bringing out this brief.

Contents

**Low Profile Conformal Antenna Arrays on High Impedance
Substrate** 1
1   Introduction 1
2   Design of High Impedance Substrate (HIS) 2
     2.1   Need of HIS 3
     2.2   Performance of HIS 4
3   Design of Dipole with High Impedance Substrate 15
     3.1   Cylindrical Dipole 15
     3.2   Printed Dipole 18
     3.3   Folded Dipole 24
4   Low Profile Dipole Antenna on Non-planar High
     Impedance Substrate 31
     4.1   Printed Dipole on a Single-Layered
            Dogbone-Based HIS 31
     4.2   Folded Dipole on Double-Layered
            Dogbone-Based HIS 36
5   Low Profile Dipole Array on Planar and Non-planar
     Dogbone-Based HIS 38
     5.1   Dogbone-Based Double-Layered Planar HIS 38
     5.2   Dogbone-Based Double-Layered Non-planar HIS 42
6   Conclusion 48
References 49

About the Book 51

Author Index 53

Subject Index 55

About the Authors

Dr. Hema Singh is currently working as Senior Scientist in Centre for Electromagnetics of CSIR-National Aerospace Laboratories, Bangalore, India. Earlier, she was Lecturer in EEE, BITS, Pilani, India during 2001–2004. She obtained her Ph.D. degree in Electronics Engineering from IIT-BHU, Varanasi India in 2000. Her active area of research is Computational Electromagnetics for Aerospace Applications. More specifically, the topics she has contributed to, are GTD/UTD, EM analysis of propagation in an indoor environment, phased arrays, conformal antennas, radar cross section (RCS) studies including Active RCS Reduction. She received Best Woman Scientist Award in CSIR-NAL, Bangalore for period of 2007–2008 for her contribution in the areas of phased antenna array, adaptive arrays, and active RCS reduction. Dr. Singh has co-authored one book, one book chapter, and over 120 scientific research papers and technical reports.

R. Chandini obtained her BE (ECE) degree from Visvesvaraya Technological University, Karnataka. She was a Project Engineer at the Centre for Electromagnetics of CSIR-National Aerospace Laboratories, Bangalore, where she worked on RCS studies and conformal arrays.

Dr. Rakesh Mohan Jha was Chief Scientist & Head, Centre for Electromagnetics, CSIR-National Aerospace Laboratories, Bangalore. Dr. Jha obtained a dual degree in BE (Hons.) EEE and M.Sc. (Hons.) Physics from BITS, Pilani (Raj.) India, in 1982. He obtained his Ph.D. (Engg.) degree from Department of Aerospace Engineering of Indian Institute of Science, Bangalore in 1989, in the area of computational electromagnetics for aerospace applications. Dr. Jha was a SERC (UK) Visiting Post-Doctoral Research Fellow at University of Oxford, Department of Engineering Science in 1991. He worked as an Alexander von Humboldt Fellow at the Institute for High-Frequency Techniques and Electronics of the University of Karlsruhe, Germany (1992–1993, 1997). He was awarded the Sir C.V. Raman Award for Aerospace Engineering for the Year 1999. Dr. Jha was elected Fellow of INAE in 2010, for his contributions to the EM Applications to Aerospace

Engineering. He was also the Fellow of IETE and Distinguished Fellow of ICCES. Dr. Jha has authored or co-authored several books, and more than five hundred scientific research papers and technical reports. He passed away during the production of this book of a cardiac arrest.

List of Figures

Figure 1   HIS unit cell consisting of paired metallic dogbones.
           a Design. **b** Dimensions   5
Figure 2   Boundary conditions imposed on HIS unit cell   5
Figure 3   Return loss at Port 2 of designed HIS unit cell consisting
           of paired PEC dogbone   6
Figure 4   Return loss at Port 2 of designed HIS unit cell consisting
           of paired copper dogbone   6
Figure 5   Single-layered *square-patch* HIS. **a** Design. **b** Excitation
           through waveguide port   7
Figure 6   Phase of reflection coefficient at the *top* of single-layered
           square-patch-based HIS   7
Figure 7   Double-layered *square-patch* HIS. **a** Model. **b** Excitation
           through waveguide port   8
Figure 8   Phase of reflection coefficient at the *top* of double-layered
           square-patch HIS.   9
Figure 9   Single-layered dogbone-based HIS. **a** Model.
           b Schematic   9
Figure 10 Phase of reflection coefficient at the *top* of single-layered
           dogbone-based HIS   10
Figure 11 HIS composed of a doubly periodic 3×3 array of metallic
           dogbones printed on PEC-backed dielectric material
           (Rogers RT 5880)   10
Figure 12 Phase of the reflection coefficient at the *top* of HIS shown
           in Fig. 11   11
Figure 13 Double-layered dogbone-based HIS. **a** Model.
           b Schematic   12
Figure 14 Phase of reflection coefficient at the *top* of double-layered
           dogbone-based HIS   12
Figure 15 HIS composed of a double layer of 3×3 array of metallic
           dogbones printed on PEC-backed dielectric material
           (Rogers RT 5880)   13

Figure 16   Phase of the reflection coefficient at the *top* of HIS shown
             in Fig. 15   13
Figure 17   HIS composed of a double layer of 3 × 3 array of metallic
             cross dogbones printed on PEC-backed dielectric material
             (Rogers RT 5880)   14
Figure 18   Phase of the reflection coefficient at the *top* of HIS shown
             in Fig. 17   14
Figure 19   A cylindrical dipole antenna in free space   15
Figure 20   A cylindrical dipole antenna on PEC   16
Figure 21   **a** A cylindrical dipole antenna on HIS consisting
             of metallic *square patches*. **b** Unit cell consisting
             of *copper square patch*.   16
Figure 22   **a** A cylindrical dipole antenna on HIS consisting
             of metallic dogbones. **b** Unit cell consisting of copper
             dogbones on dielectric substrate.   17
Figure 23   Comparison of return loss of cylindrical dipole
             over different substrates   18
Figure 24   **a** Schematic of printed dipole. **b** A printed dipole antenna
             on HIS consisting of metallic *square patches*. **c** Return
             loss of a printed dipole antenna on HIS consisting of
             metallic *square patches*   19
Figure 25   **a** Return loss of a printed dipole on a single-layered
             square-patch HIS. **b** Radiation pattern of a printed dipole
             on a single-layered *square-patch*-based HIS   20
Figure 26   **a** Printed dipole on a double-layered *square-patch* HIS.
             b Return loss of a printed dipole on a double-layered
             square-patch HIS. **c** Radiation pattern of a printed dipole
             on a double-layered *square-patch*-based HIS   22
Figure 27   A printed dipole antenna on HIS consisting of copper
             dogbones   23
Figure 28   Return loss of a printed dipole antenna on HIS consisting
             of copper dogbones   23
Figure 29   Gain of a printed dipole antenna on HIS consisting
             of copper dogbones   24
Figure 30   A printed dipole on double-layered copper dogbone-based
             HIS. **a** Parallel dogbones. **b** Crossed dogbones; Dipole
             aligned along the upper dogbone layer. **c** Crossed
             dogbones; Dipole aligned along the lower dogbone
             layer   25
Figure 31   Return loss of double-layered dogbone-based HIS
             consisting of **a** Parallel dogbones. **b** Crossed dogbones;
             dipole aligned along the *upper* dogbone layer.
             c Crossed dogbones; dipole aligned along the *lower*
             dogbone layer   26

Figure 32   Gain of printed dipole antenna on cross double dogbone
             layer HIS (Fig. 30c).   27
Figure 33   Return loss of printed dipole antenna on double layer
             of dogbone-based HIS (Fig. 30c) in THz region   27
Figure 34   A folded dipole antenna on HIS consisting of copper
             dogbones   28
Figure 35   Return loss of folded dipole antenna on HIS consisting
             of copper dogbones. **a** Dipole substrate's $\varepsilon_r = 2.33$,
             $\tan \delta_e = 0.0012$. **b** Dipole substrate's $\varepsilon_r = 2.2$,
             $\tan \delta_e = 0.0009$ 29
Figure 36   Gain of folded dipole antenna on HIS (Fig. 35b).   30
Figure 37   Folded dipole antenna on cross double dogbone layer
             HIS   30
Figure 38   Return loss of a folded dipole antenna on cross double
             dogbone layer HIS. **a** Dipole substrate's $\varepsilon_r = 3$, $\tan \delta_e = $
             0.0013. **b** Dipole substrate's $\varepsilon_r = 4.5$, $\tan \delta_e = 0.002$ 32
Figure 39   Return loss of a folded dipole antenna on cross double
             dogbone layer HIS with dipole substrate ($\varepsilon_r = 3$, $\tan \delta_e = $
             0.0013). **a** Dogbone substrate's $\varepsilon_r = 6.15$, $\tan \delta_e = 0.0025$.
             b Dogbone substrate's $\varepsilon_r = 9.2$, $\tan \delta_e = 0.0023$.   33
Figure 40   Gain of folded dipole antenna on double-layered cross
             dogbone-based HIS (Fig. 38b).   34
Figure 41   Folded antenna on HIS consisting of double layer
             of parallel dogbones.   34
Figure 42   Return loss folded dipole antenna on HIS consisting
             of copper dogbones. **a** Dipole substrate's $\varepsilon_r = 3$, $\tan \delta_e = $
             0.0013. **b** Dipole substrate's $\varepsilon_r = 4.5$, $\tan \delta_e = 0.002$ 35
Figure 43   Single-layered dogbone-based non-planar HIS   36
Figure 44   Phase of reflection coefficient at the top of single-layered
             dogbone-based non-planar HIS   36
Figure 45   Printed dipole on a single-layered non-planar dogbone
             HIS   37
Figure 46   Return loss of a printed dipole on a single-layered dogbone
             non-planar HIS   37
Figure 47   Radiation pattern of a printed dipole on a single-layered
             dogbone-based non-planar HIS   38
Figure 48   Double-layered dogbone-based non-planar HIS. **a** Model.
             b Schematic   39
Figure 49   Phase of reflection coefficient at the top of double-layered
             dogbone-based non-planar HIS   40
Figure 50   Folded dipole on a double-layered dogbone-based
             non-planar HIS   40
Figure 51   Return loss of a folded dipole on a double-layered
             dogbone curved HIS   41

Figure 52   Radiation pattern of a folded dipole on a double-layered
             dogbone-based non-planar HIS   41
Figure 53   2-Element printed dipole array on a double-layered
             dogbone-based planar HIS   42
Figure 54   Return loss of a 2-element printed dipole array
             on a double-layered dogbone-based planar HIS   42
Figure 55   Radiation pattern of a 2-element printed dipole array
             on a double-layered dogbone-based planar HIS   43
Figure 56   3-Element printed dipole array on a double-layered
             dogbone-based planar HIS   43
Figure 57   Return loss of a 3-element printed dipole array
             on a double-layered dogbone-based planar HIS   44
Figure 58   Radiation pattern of a 3-element printed dipole array
             on a double-layered dogbone-based planar HIS   44
Figure 59   2-Element folded dipole array on a double-layered
             dogbone-based non-planar HIS   45
Figure 60   Return loss of a 2-element folded dipole array
             on a double-layered dogbone-based non-planar HIS   46
Figure 61   Radiation pattern of a 2-element folded dipole array
             on a double-layered dogbone-based non-planar HIS   46
Figure 62   3-Element folded dipole array on a double-layered
             dogbone-based non-planar HIS   47
Figure 63   Return loss of a 3-element folded dipole array
             on a double-layered dogbone-based non-planar HIS   47
Figure 64   Radiation pattern of a 3-element folded dipole array
             on a double-layered dogbone-based non-planar HIS   48

Low Profile Conformal Antenna Arrays on High Impedance Substrate

Abstract The High Impedance Surface (HIS) is a preferred substrate for low profile antenna design, owing to its unique boundary conditions. Such substrates permit radiating elements to be printed on them, without any disturbance in the radiation characteristics. Moreover, HIS provides improved impedance matching, enhanced bandwidth, and increased broadside directivity owing to total reflection from the reactive surface and high input impedance. This book presents electromagnetic (EM) design and analysis of dipole antenna array over the high impedance substrate. Different configurations of HIS are considered for the array design on planar and non-planar high-impedance surface. Results are presented for cylindrical dipole, printed dipole, and folded dipole over single and double layered square-patch based-HIS and dogbone-based HIS. The performance of antenna array is analyzed in terms of performance parameters such as return loss and radiation pattern. The design presented shows acceptable return loss and mainlobe gain of radiation pattern. This book provides insight into the EM design and analysis of conformal arrays. The book is a comprehensive text for beginners in the design and analysis of HIS-based antenna arrays. It includes pictorial descriptions of both planar and non-planar array design and detailed discussion of the performance analysis of HIS-based planar and non-planar antenna arrays.

Keywords High impedance substrate · Cylindrical dipole · Printed dipole · Folded dipole · Dogbone

1 Introduction

The electromagnetic bandgap (EBG) material and metamaterials have been used for the development of artificial magnetic conductors (AMC) or high impedance surface (HIS). These surfaces serve as an effective shielding to low profile electric dipole antenna, without disturbing the associated tangential electric field (Best and Hanna 2008; Vallecchi and Capolino 2009a, b).

© The Author(s) 2016   1
H. Singh et al., *Low Profile Conformal Antenna Arrays*
on High Impedance Substrate, SpringerBriefs in Computational Electromagnetics,
DOI 10.1007/978-981-287-763-5_1

In a conventional antenna design, the conducting ground planes are used in the vicinity of the antenna for confining the radiation within the half-space. However, due to small distance between the radiating element and ground plane, the destructive interference between the image currents and the radiating currents take place, reducing the efficiency of the antenna. On the other hand, HIS support in-phase image currents facilitates the constructive interference of the radiating fields. This results in enhanced directivity of the antenna. The reflection coefficient of HIS is almost +1 at frequencies, where the surface acts like AMC and −1 at frequencies where the surface acts like perfect electric conductor (PEC) (McVay et al. 2003). The low profile antennas, viz. microstrip patch antennas, printed dipoles, folded dipoles show better performance if their low impedance ground plane is replaced by HIS. The HIS can be planar or non-planar (Colburn et al. 2007). Conformal antennas are preferred in mobile and commercial communications and military communication systems due to their wide angle coverage, and aerodynamic compatibility to the platform (Erturk and Guner 2004). However, the shortage of efficient and accurate tools for design and analysis of conformal printed antenna arrays, in particular for electrically large ones, makes this problem an important one.

In this book, attempt has been made to design low profile dipole array on a high impedance substrate consisting of array of metallic dogbones or square patches. The HIS unit cell design consists of single- and double-layered conductors over a PEC ground plane-backed dielectric substrate. The antenna elements are placed on a top surface of metamaterial/HIS consisting of tightly coupled metallic dogbones. The substrate is backed by PEC ground plane. The characteristics of dogbone-based HIS is compared with that of square-patch HIS. The low profile antennas, viz. cylindrical dipole, printed dipole, and folded dipole antenna are designed over a single- and double-layered dogbone-based HIS. Two configurations of double-layer dogbone-based HIS is considered. These include double layer of parallel metallic dogbones and double layer of perpendicular (cross) metallic dogbones. The total antenna design along with multilayered substrate and PEC ground plane is within subwavelengths. The role of dipole element and dogbone substrates is analyzed on the antenna performance. The performance of antenna array is studied in terms of amplitude, phase of reflection coefficient (i.e., return loss), and the radiation pattern. It is shown that the multilayered HIS contributes toward the wider fractional bandwidth of the antenna, better impedance matching. The overall design performance depends on both the radiating element and the HIS parameters. The array design is extended for non-planar HIS. The study is geared toward the design and analysis of conformal antenna array over high impedance substrates.

2 Design of High Impedance Substrate (HIS)

The HIS is designed using tightly coupled metallic conductors over a planar dielectric substrate (Donzelli et al. 2009; Vallecchi et al. 2012). These conductors can be either square patches (Mosallaei and Sarabandi 2004; Cure et al. 2013), or mushroom-like structure (Sievenpiper et al. 1999; Azad and Ali 2008) or dogbones (Vallecchi and Capolino 2009a, b).

2.1   Need of HIS

The characteristics of a dipole in a free space are altogether different, if it is placed near a conducting or dielectric surface. The directivity of a dipole over a PEC surface is better than for a perfect magnetic conductor (PMC) case. If the distance between the dipole and the PEC surface is decreased, the input resistance falls drastically because of cancelation of radiating and image currents. In contrast, for PMC, the input resistance increases owing to the constructive interference of radiating and image currents (Vallecchi and Capolino 2009a, b). The lower radiation resistance implies low radiation efficiency. Moreover, the impedance matching for the antenna placed close to the PEC ground plane is difficult. In case of PMC. the impedance matching for low profile dipole is comparatively easier. If dipole is placed at distance greater than half-wavelength from the PEC or PMC surface, the input impedance increases to that of free space impedance. The directivity of the dipole antenna is higher in case of PEC than PMC, especially for smaller distance.

Another convention of designing low profile antenna is to place it over PEC backed dielectric slab. The high permittivity of the substrate allows the miniaturization of the radiating element, but simultaneously affects the radiation behavior of the antenna. It supports the surface waves, trapping significant amount of radiated field, and hence distort the antenna pattern.

In order to increase the radiation efficiency of low profile antenna, the substrate is modified by including tightly coupled metallic conductors embedded in a dielectric. Such HIS provides high input impedance to the dipole printed over it. The high impedance of the substrate allows the antenna to be printed close to the surface without any distortion to its radiation pattern. This improves the gain performance of the antenna printed over the HIS. Further, the substrate blocks the radiation to travel across thereby increasing the antenna efficiency and ease of low profile miniaturized antenna design (Vallecchi et al. 2012). Table 1 summarizes the performance of PEC, PMC, dielectric substrate, and HIS for low profile antenna design. The metamaterials made up of paired dogbones (Donzelli et al. 2009) exhibits both electric and magnetic resonance. Their design give rise to symmetric and antisymmetric current distribution on top and bottom conducting surface of each dogbone pair respectively. Moreover, the HIS supports the surface and leaky waves, and hence contributing toward the enhancement of antenna performance (Vallecchi and Capolino 2009a, b).

Table 1 Characteristics of surface for low profile antenna design

Surface	Advantage	Disadvantage
PEC	High broadside directivity	Difficulty in impedance matching
	High quality factor	Narrow impedance bandwidth
		Low input resistance for antenna close to PEC surface
PMC	Improved bandwidth	Low directivity
	Better impedance matching	
	Higher input resistance for antenna close to PMC surface	
Dielectric substrate	Support antenna miniaturization	Supports surface wave
	Moderate input resistance	High permittivity results in pattern distortion
HIS	Improved bandwidth with respect to PEC	Complexity in design
	High directivity with respect to PMC	
	High input impedance	
	Improved antenna efficiency	
	Support antenna miniaturization	

2.2 Performance of HIS

The performance of HIS may be analyzed in terms of parameters like magnitude and phase of the reflection/transmission coefficient of the surface, surface reactance etc. The magnitude of the reflection coefficient, commonly referred to as return loss ($=10 \log_{10} \Gamma$) quantifies the dielectric and conductor losses. This return loss is expected to be very low (less than -20 dB) for HIS consisting of patterned conductors embedded within the thin dielectric substrate toward compact low profile antenna design.

2.2.1 Design of HIS Unit Cell

The HIS is designed using an array of pairs of metallic dogbone-shaped conductors. The dogbone pair is printed on a dielectric substrate (Rogers RO3003: $\varepsilon_r = 3$ and $\tan \delta_e = 0.0013$) with thickness, $H = 0.02\lambda$. The dogbone material is taken as PEC with thickness, $T = 6.416 \times 10^{-3}\lambda$. The unit cell of the HIS (Fig. 1) is designed with the following geometric parameters; $A = 0.126\lambda$, $B = 0.126\lambda$, $A1 = 0.014\lambda$, $A2 = 0.123\lambda$, $B1 = 0.062\lambda$, and $B2 = 0.0126\lambda$.

The unit cell is excited using waveguide ports ($z = -H$ and $z = 2H + T$). The boundary conditions imposed on the unit cell are shown in Fig. 2. The resultant return loss of the unit cell with PEC dogbone pair is shown in Fig. 3. In practical designs, instead of PEC a lossy metal is preferred. The corresponding return loss of

Fig. 1 HIS unit cell consisting of paired metallic dogbones. **a** Design. **b** Dimensions

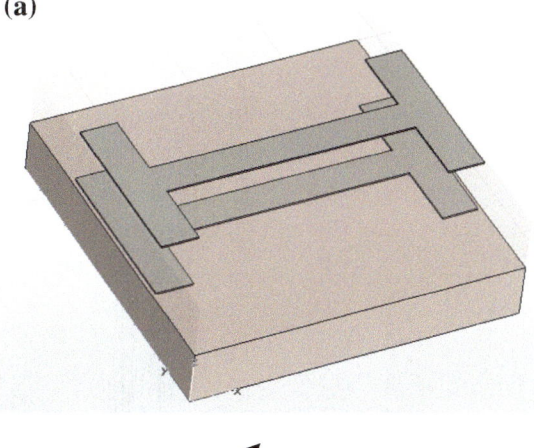

(a)

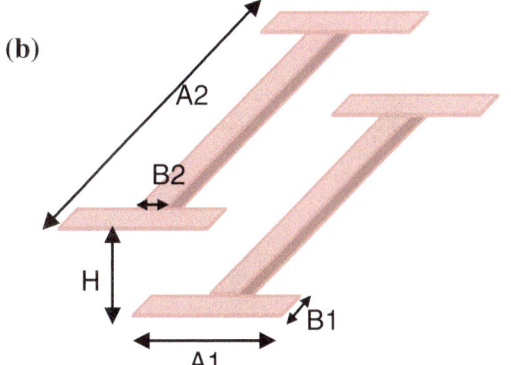

(b)

Fig. 2 Boundary conditions imposed on HIS unit cell

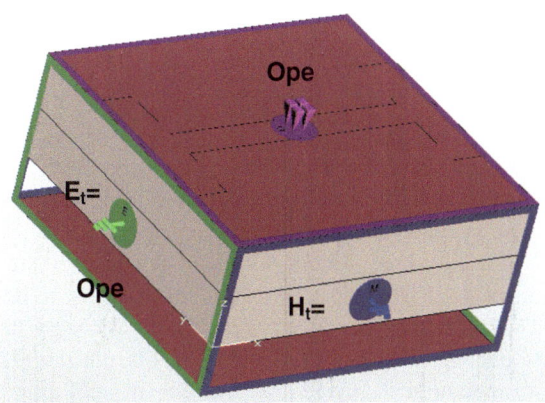

unit cell with copper dogbones is shown in Fig. 4. The lossy nature of copper as compared to PEC is evident from the change in the depth of return loss. However, the resonant frequency of 5.3 GHz remains unchanged.

Fig. 3 Return loss at Port 2
of designed HIS unit cell
consisting of paired PEC
dogbone

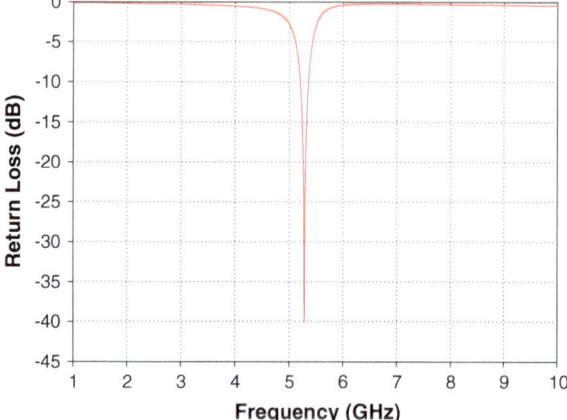

Fig. 4 Return loss at Port 2
of designed HIS unit cell
consisting of paired copper
dogbone

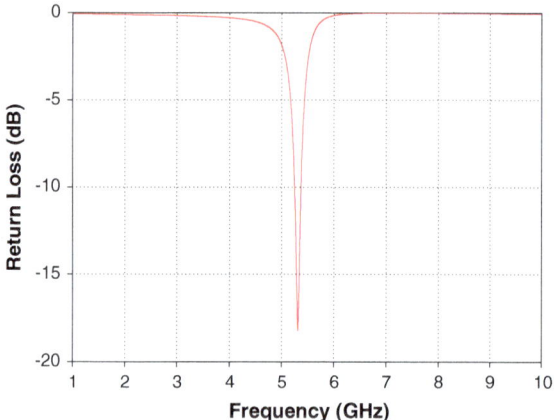

2.2.2 Design of Planar HIS Substrate

This section presents the EM design and analysis of planar HIS. First, the
high-impedance substrate is designed using metallic square patches and then based
on metallic dogbones. Both single- and double-layered HIS are studied in terms of
phase of reflection coefficient.

Single-layered square-patch-based HIS: Figure 5a shows the design of a
square patch HIS on a dielectric substrate ($\varepsilon_r = 10.2$, $\tan \delta = 0.0035$) of height,
$H = 1.61$ mm. The square patch is taken as copper ($\sigma = 5.8 \times 10^7$ S/m). The design
parameters of square patch are $L_s = 4.6$ mm; $t = 0.035$ mm; gap = 2.4 mm. The
substrate dimensions are taken as $L = W = 24$ mm.

The excitation is done through waveguide port, as shown in Fig. 5b. The phase
of reflection coefficient is shown in Fig. 6. The substrate resonates at 9.3 GHz.

Double-layered square-patch-based HIS: Next a double-layered square
patch-based HIS is designed ($\varepsilon_r = 10.2$, $\tan \delta = 0.0035$) of height, $H = 1.61$ mm

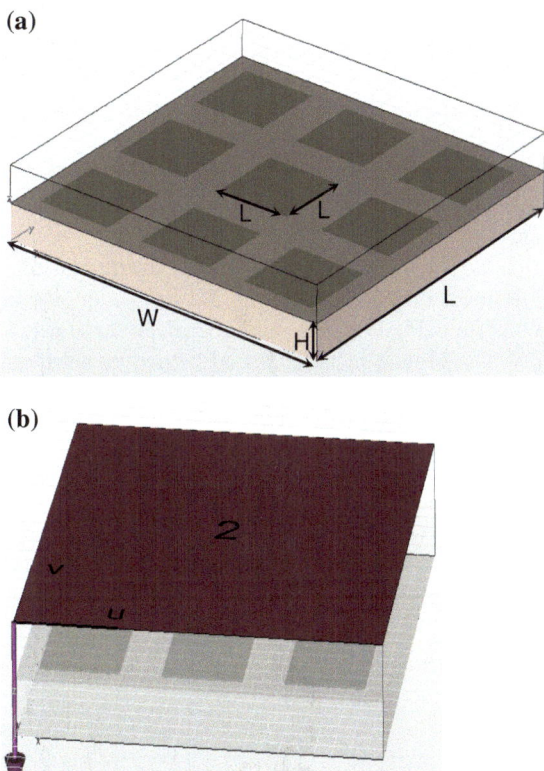

Fig. 5 Single-layered *square-patch* HIS. **a** Design. **b** Excitation through waveguide port

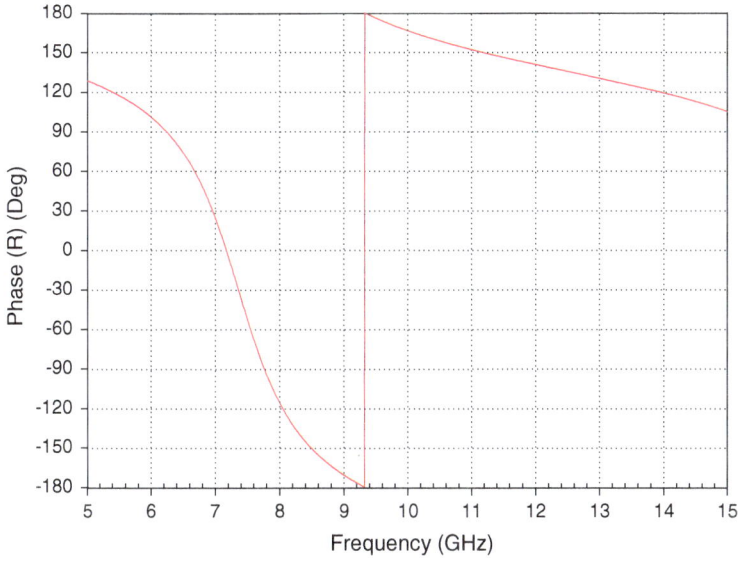

Fig. 6 Phase of reflection coefficient at the *top* of single-layered *square-patch*-based HIS

(Fig. 7a). The design parameters of square patch are $L_s = 4.6$ mm; $t = 0.035$ mm; gap $= 2.4$ mm. The substrate dimensions are $L = W = 24$ mm. The excitation is done through waveguide port which is placed on top of the HIS (Fig. 7b). The phase of reflection coefficient is shown in Fig. 8. As expected the double-layered substrate resonates twice (6.3 and 9.5 GHz).

Single-layered dogbone-based HIS: The array of metallic dogbone-shaped conductor imposes high impedance to the radiating element printed over it. A single-layered copper dogbone-based HIS (Fig. 9a) is designed with dielectric substrate ($\varepsilon_r = 10.2$, $\tan \delta = 0.0035$; $H = 1.61$ mm). The dimension of dogbones are $A1 = 0.875$ mm, $B1 = 3.5$ mm, $A2 = 6.83$ mm, $B2 = 0.7$ mm, gap $= 1.045$ mm, and thickness, $T = 0.035$ mm (Fig. 9b). Excitation is given through waveguide port at the top of the HIS. The ground plane is placed below the substrate.

The phase of the reflection coefficient of the design is shown in Fig. 10. The resonant frequency is observed as 4.8 GHz. Next another configuration of single-layered metallic dogbones over a dielectric substrate is designed. The 3×3 doubly periodic metallic dogbones printed on a dielectric substrate (Rogers RT 5880 with permittivity, $\varepsilon_r = 2.2$ and loss tangent, $\tan \delta_e = 0.0009$) of thickness,

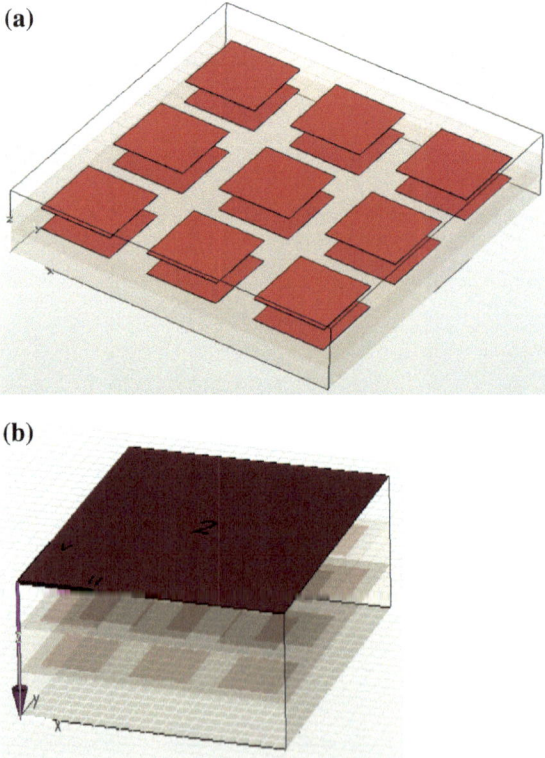

(a)

(b)

Fig. 7 Double-layered *square-patch* HIS. **a** Model. **b** Excitation through waveguide port

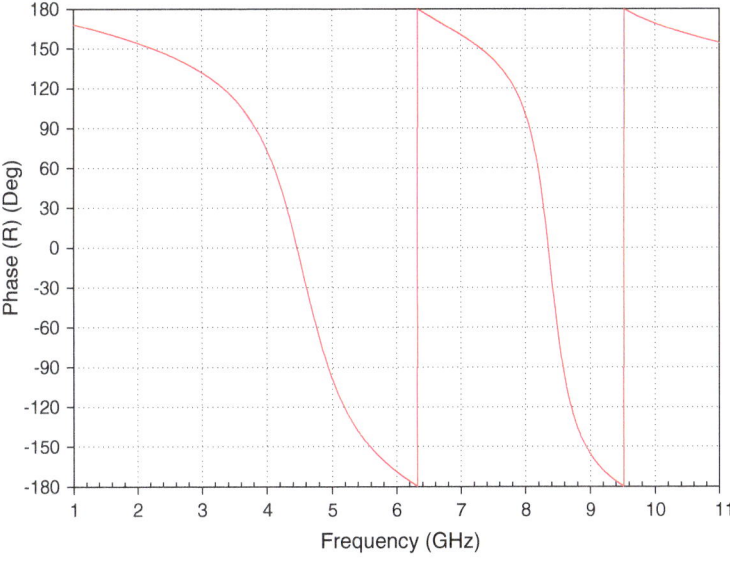

Fig. 8 Phase of reflection coefficient at the *top* of double-layered *square-patch* HIS

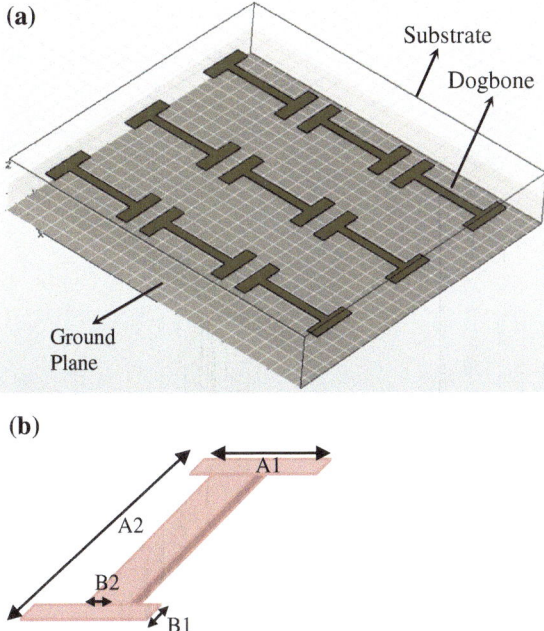

Fig. 9 Single-layered dogbone-based HIS. **a** Model. **b** Schematic

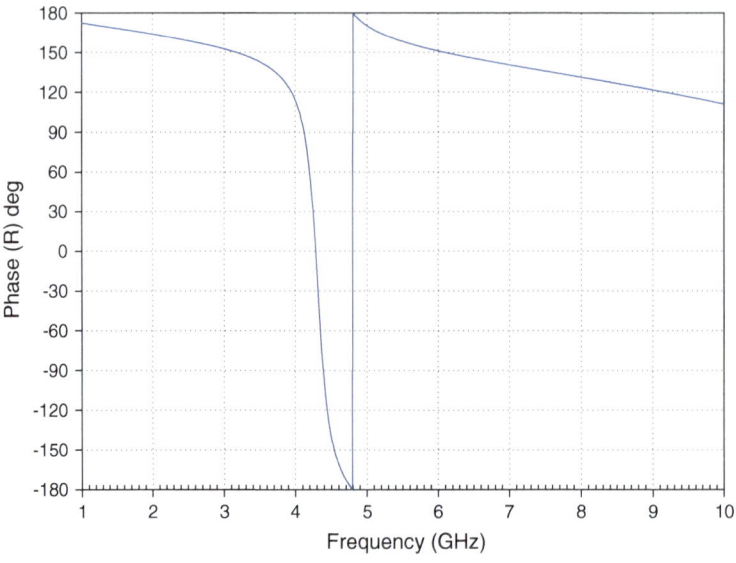

Fig. 10 Phase of reflection coefficient at the *top* of single-layered dogbone-based HIS

$H = 1.61$ mm is shown in Fig. 11. The dogbone material is taken as copper with thickness, $T = 0.035$ mm. The geometric parameters of the model are $A = 7$ mm, $B = 7$ mm, $A1 = 0.875$ mm, $A2 = 6.83$ mm, $B1 = 3.5$ mm, and $B2 = 0.7$ mm. The design frequency is taken as 5.5 GHz. The single waveguide port is placed at $2H + T$ from the bottom of the ground plane.

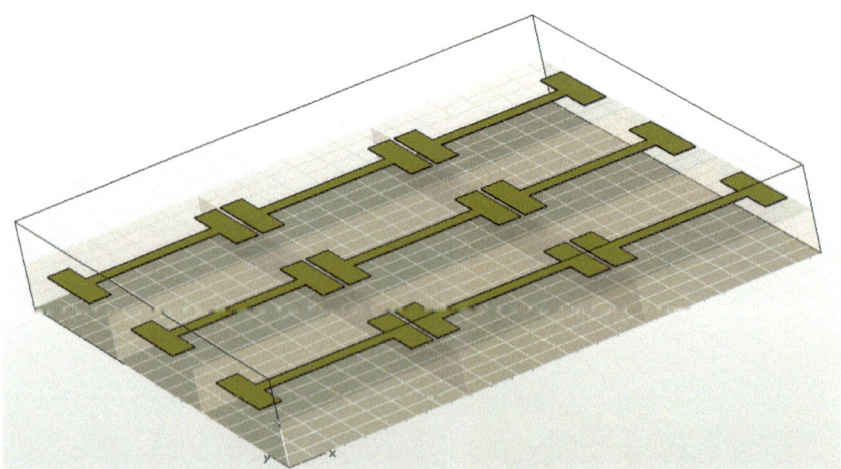

Fig. 11 HIS composed of a doubly periodic 3 × 3 array of metallic dogbones printed on PEC-backed dielectric material (Rogers RT 5880)

The phase of the reflection coefficient at the port is shown in Fig. 12. The bandwidth achieved is 20 %. The magnetic resonance takes place, where phase of the reflection coefficient approaches zero. At this point, the surface will act as AMC, imposing high input impedance.

From Fig. 12, it may be observed that phase of reflection coefficient goes to zero at f_m = 5.6 GHz. The surface will be inductive and capacitive below and above f_m respectively (Vallecchi et al. 2012). Further, the surface will act as PEC when the phase tends to ±180°. The location of magnetic resonance is controlled by design parameters like dogbone dimensions, constitutive parameters of substrate (Donzelli et al. 2009).

For the states in between AMC and PEC, the substrate exhibits either inductive or capacitive characteristics. The substrate offers high impedance near resonance frequency. If the radiating element like dipole is placed above such substrate, high efficiency, and better impedance matching may be achieved.

Double-layered dogbone-based HIS: Next single-layered dogbone-based HIS is modified into double-layered dogbone-based HIS (Fig. 13), keeping the dogbone parameters same as that of Fig. 9b.

The permittivity and loss tangent of the HIS substrate are taken as 4.5 and 0.002 respectively. Waveguide port is used for the excitation. Figure 14 presents the phase of the reflection coefficient of the design. The resonant frequencies are observed as 5.25 and 6.8 GHz.

Next, another configuration of double-layered metallic dogbones is designed over the PEC-backed dielectric substrate (Rogers RT 5880 with permittivity, ε_r = 2.2; $\tan \delta_e$ = 0.0009; H = 1.61 mm). The thickness of first and second layer

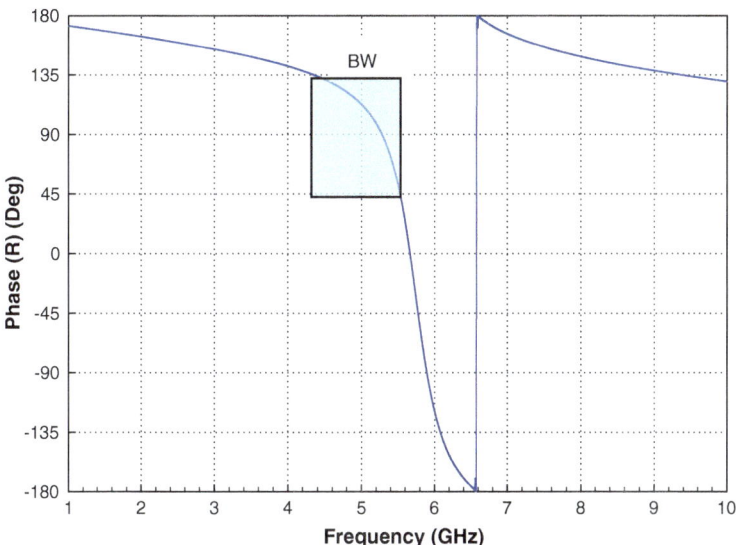

Fig. 12 Phase of the reflection coefficient at the *top* of HIS shown in Fig. 11

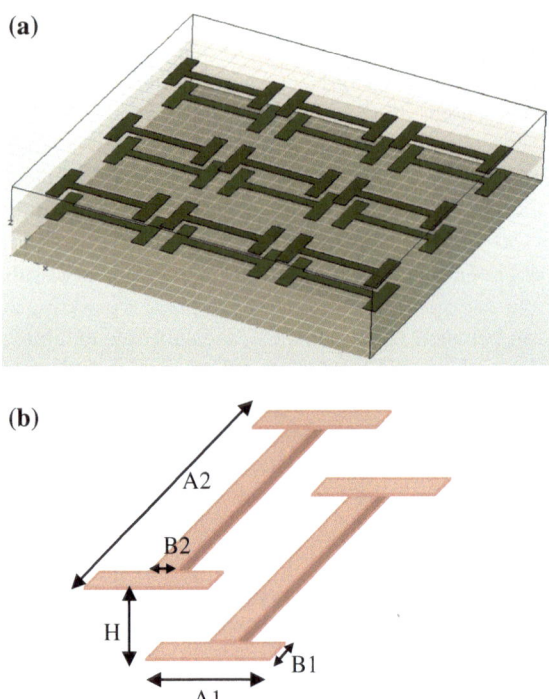

Fig. 13 Double-layered dogbone-based HIS. **a** Model. **b** Schematic

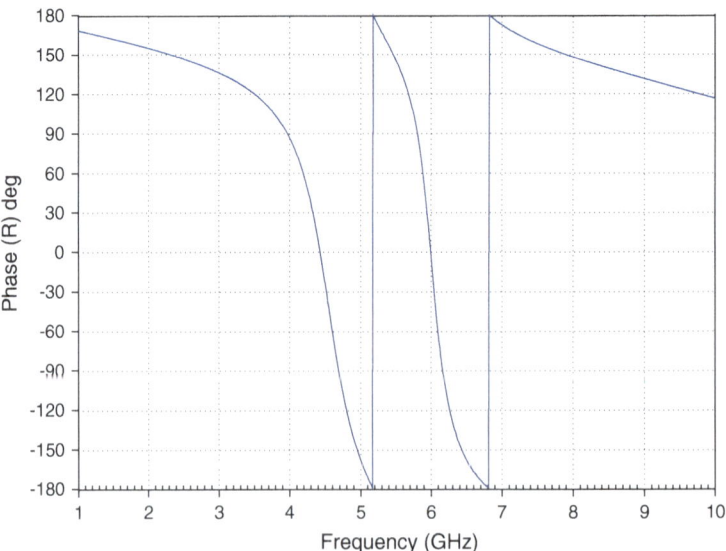

Fig. 14 Phase of reflection coefficient at the *top* of double-layered dogbone-based HIS

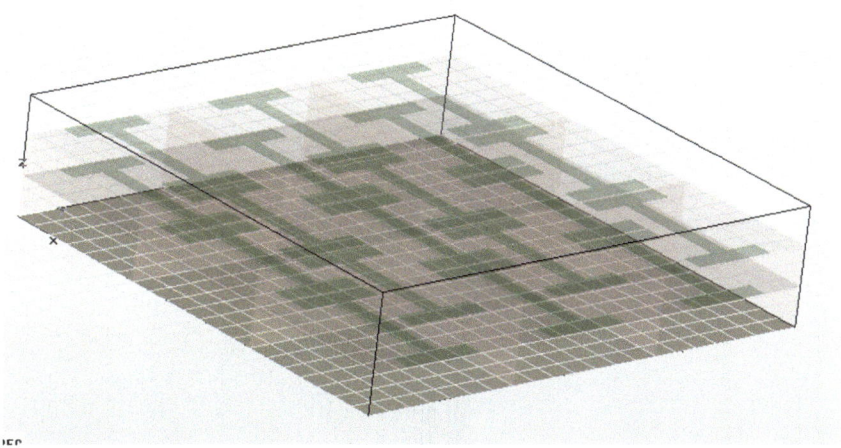

Fig. 15 HIS composed of a double layer of 3 × 3 array of metallic dogbones printed on PEC-backed dielectric material (Rogers RT 5880)

substrate is kept same. The dogbones dimensions are same as that of single-layer substrate (Fig. 5). The total thickness of the double-layer HIS (Fig. 15) is $2H + T$ (3.255 mm). The excitation is given through waveguide port at $z = 4.79$ mm. The phase of the reflection coefficient at the port is shown in Fig. 16. The phase of the reflection coefficient becomes 180° at two frequencies (5.2 and 6.8 GHz), owing to the two layers of dogbones in HIS.

There can be other configuration consisting of cross dogbones in HIS (Fig. 17). The dogbones of first and second layer are perpendicular to each other. The

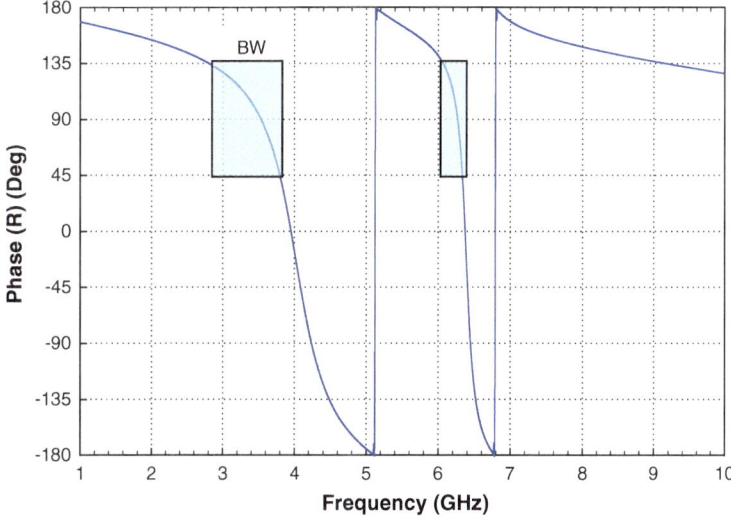

Fig. 16 Phase of the reflection coefficient at the *top* of HIS shown in Fig. 15

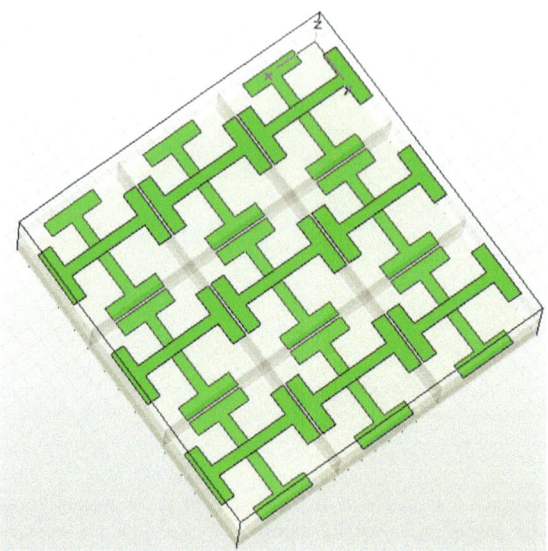

Fig. 17 HIS composed of a double layer of 3 × 3 array of metallic cross dogbones printed on PEC-backed dielectric material (Rogers RT 5880)

dimension of dogbones and substrate are kept same as in Fig. 15. The corresponding phase of the reflection coefficient is shown in Fig. 18. It may be seen that the phase of the reflection coefficient reaches at 180° only at one frequency (6.2 GHz) similar to Fig. 12 and the bandwidth is enhanced to 23 % as compared to Fig. 12.

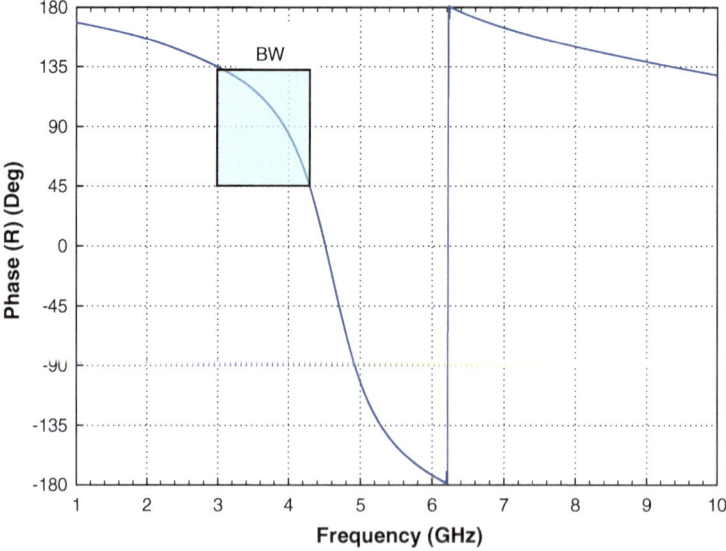

Fig. 18 Phase of the reflection coefficient at the *top* of HIS shown in Fig. 17

3 Design of Dipole with High Impedance Substrate

Once the substrate is designed, a metallic dipole is placed on the top of it. The simulations are performed for normal incidence and linear polarization. The center frequency for the design is taken as 5.5 GHz. It is known that the dipole exhibits capacitive behavior below its resonant frequency (Kraus et al. 2006). As stated earlier, the HIS is inductive below magnetic resonance. If the dipole is placed over HIS, the inductive behavior, i.e., ability to store magnetic energy would compensate for stored electric energy associated with the dipole below its resonant frequency (Vallecchi et al. 2012). In other words, the dipole would resonate at frequency lower than its resonant frequency in free space. This would reduce the antenna dimensions in terms of wavelength, contributing toward antenna miniaturization.

3.1 Cylindrical Dipole

First a cylindrical dipole is considered. The antenna substrate is an important factor in the resultant radiation pattern and return loss characteristics. Figures 19 and 20 show a cylindrical dipole in free space and on PEC surface respectively. The radius and length of dipole is taken as 0.046λ and 0.46λ respectively. The antenna is excited using discrete port. In order to analyze the performance of a cylindrical dipole antenna with different substrates, the dipole is designed with substrates like PEC, square-patch-based HIS, and dogbone-based HIS. As discussed above, PEC or PMC surface are not preferable in view of directivity and impedance matching.

3.1.1 HIS—Square Patches on PEC-Backed Dielectric Substrate

The design of HIS consisting of copper square patches is shown in Fig. 21. The permittivity and loss tangent of the dielectric substrate are taken as $\varepsilon_r = 11.5$ and $\tan\delta_e = 0.004$ respectively. The thickness of copper patch is taken as 0.035 cm. The

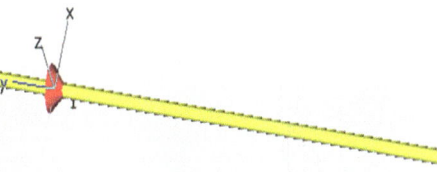

Fig. 19 A cylindrical dipole antenna in free space

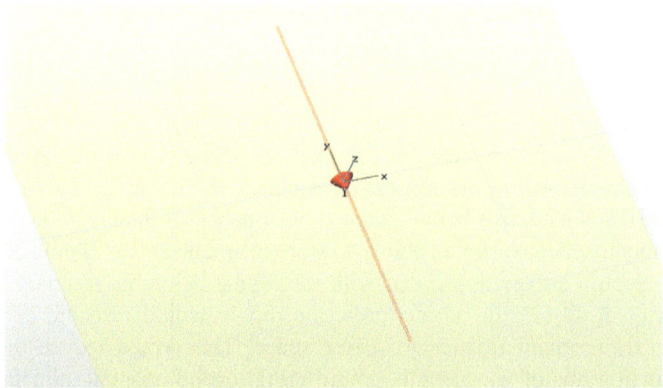

Fig. 20 A cylindrical dipole antenna on PEC

(a)

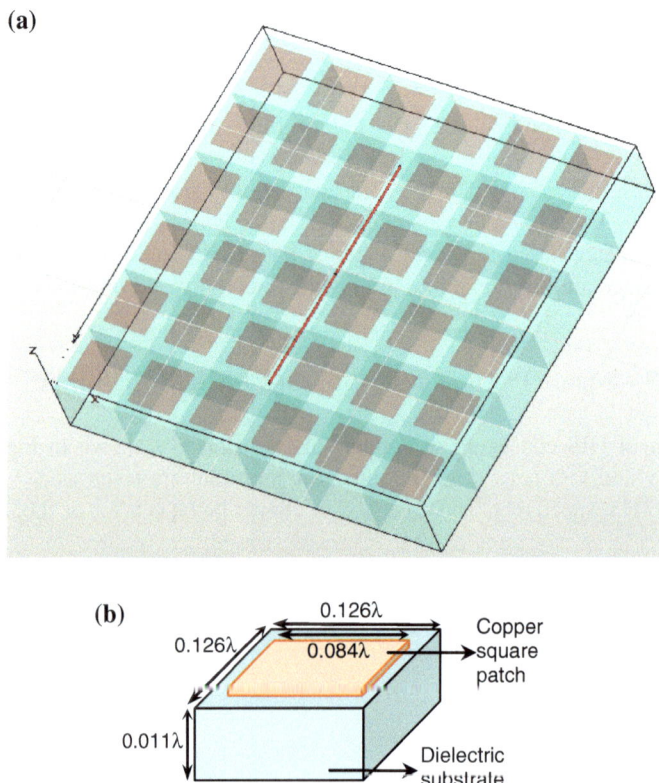

Fig. 21 a A cylindrical dipole antenna on HIS consisting of metallic *square patches*. **b** Unit cell consisting of *copper square patch*

design frequency is 5.5 GHz. A cylindrical dipole is placed on a PEC-backed HIS at $z = 0.021\lambda$, as shown in Fig. 21. The dimensions of copper patch and the dielectric substrate are shown in Fig. 21b. The dimension of dipole is same as in Fig. 19.

3.1.2 HIS—Dogbones on PEC-Backed Substrate

Another HIS consisting of doubly periodic layer of metallic dogbones is designed. As above, a cylindrical dipole is placed on PEC-backed HIS at $z = 0.021\lambda$, as shown in Fig. 22a. The dimensions of copper dogbone and substrate are shown in Fig. 22b, where $A = 0.126\lambda$, $B = 0.126\lambda$, $A1 = 0.014\lambda$, $A2 = 0.123\lambda$, $B1 = 0.062\lambda$, $B2 = 0.0126\lambda$ and $H = 0.03\lambda$. The thickness of dogbone is taken as 0.35 cm. The permittivity and loss tangent of the dielectric substrate are taken as $\varepsilon_r = 4.9$ and $\tan\delta_e = 0.025$

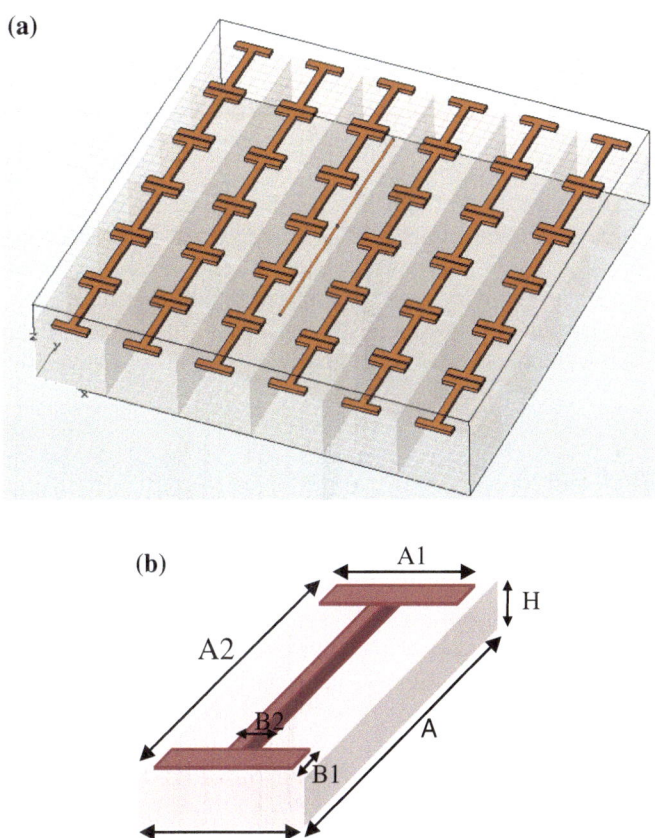

Fig. 22 **a** A cylindrical dipole antenna on HIS consisting of metallic dogbones. **b** Unit cell consisting of copper dogbones on dielectric substrate

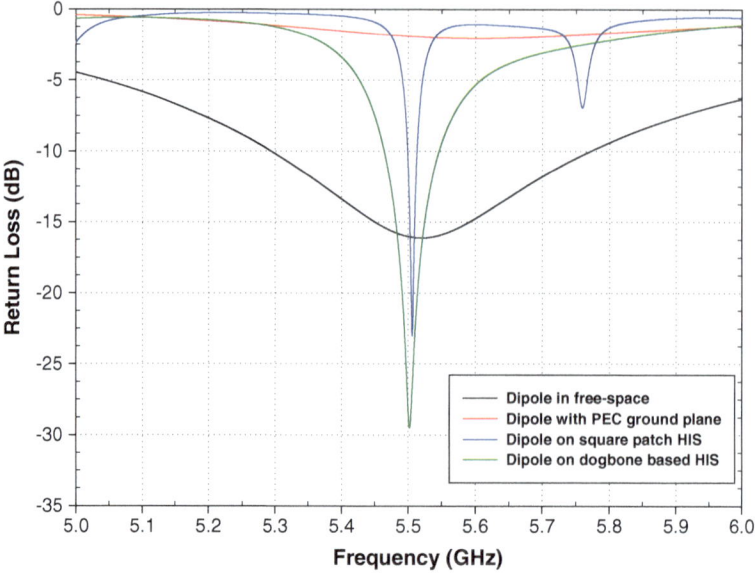

Fig. 23 Comparison of return loss of cylindrical dipole over different substrates

respectively. The return loss of a cylindrical dipole antenna is compared for free space and different substrates in Fig. 23. It is apparent that the performance of cylindrical dipole is best in terms of return loss when it is placed over based HIS.

This provides an improved impedance matching and hence high antenna efficiency. Further, it is not only good return loss but also wider bandwidth (at −10 dB) that is achieved with dogbone-based HIS, as compared to square-patch-based HIS.

3.2 Printed Dipole

For printed dipole over the HIS, an additional dielectric layer is included above the HIS. The dipole is excited through a waveguide port through twin line (Fig. 24a). The dipole antenna is designed with the length of 0.46λ (including gap) and width of 0.04λ. The twin line consists of two strips each with width of 0.01λ. The gap between the strips is 0.2 mm. The design frequency is 5.5 GHz.

3.2.1 HIS: Square Patches on PEC-Backed Substrate

The printed dipole over the HIS consisting of copper square patches is shown in Fig. 24b. The dipole substrate (upper substrate) is FR4 ($\varepsilon_r = 4.9$, $\tan \delta_e = 0.025$, $d = 0.025\lambda$) and the lower substrate ($d = 0.03\lambda$) on which the square patches are placed is of permittivity $\varepsilon_r = 7.8$ and $\tan \delta_e = 0.0015$. The dimensions of the square

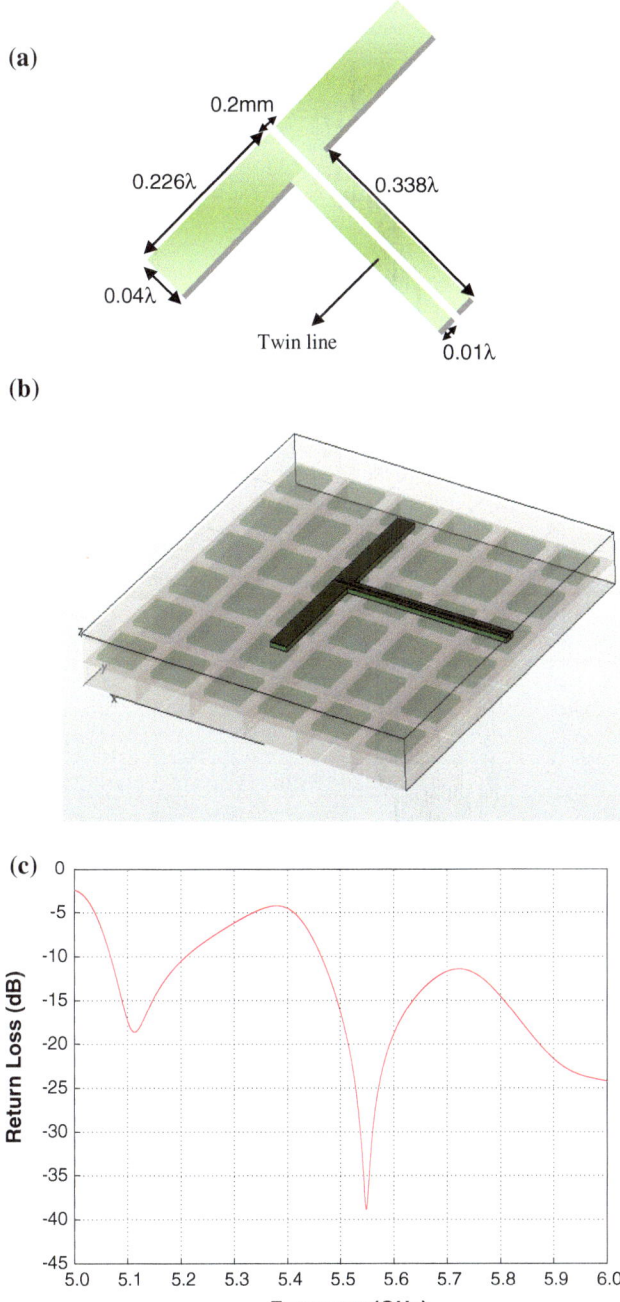

Fig. 24 **a** Schematic of printed dipole. **b** A printed dipole antenna on HIS consisting of metallic *square patches*. **c** Return loss of a printed dipole antenna on HIS consisting of metallic *square patches*

patches are same as in Fig. 21b. The dipole resonates at 5.55 GHz with return loss of −39 dB (Fig. 24c).

Next a printed dipole is designed at 10 GHz, with design parameters $L = 0.46\lambda$, $W = 0.04\lambda$, $T = 0.035$ mm. The dipole substrate has $\varepsilon_r = 4.5$, $\tan \delta = 0.002$. The HIS substrate is dielectric ($\varepsilon_r = 6.15$, $\tan \delta = 0.0019$ and $H = 0.0537\lambda$). The dipole excitation is given by waveguide port through twin line (length $= 0.338\lambda$; width $= 0.01\lambda$). Figure 25a shows that the dipole resonates at 10.5 GHz. The corresponding radiation pattern is shown in Fig. 25b. The mainlobe gain of the design is obtained as −7.5 dBi.

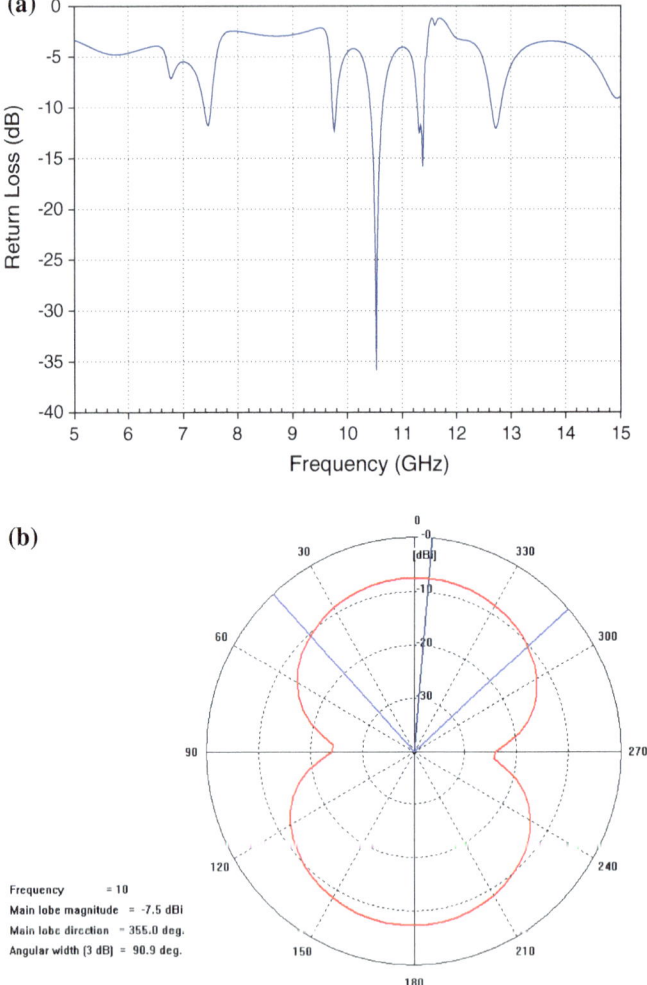

Fig. 25 **a** Return loss of a printed dipole on a single-layered *square-patch* HIS. **b** Radiation pattern of a printed dipole on a single-layered *square-patch*-based HIS

3.2.2 Double-Layered Square-Patch-Based HIS

Next, a dipole is printed on double-layered square-patch-based HIS (Fig. 26a). The design parameters including the substrates are kept same as above. The dipole substrate is taken as dielectric (ε_r = 4.5 and tan δ = 0.002; height = 0.0269λ). The waveguide excitation is given through twin-line. The return loss obtained at 10.2 GHz is −27 dB, shown in Fig. 26b. The corresponding radiation pattern is shown in Fig. 26c.

The mainlobe gain obtained is −2.8 dBi, which is higher than Fig. 25b; dipole over single-layered square-patch-based HIS.

3.2.3 HIS: Single-Layered Dogbones on PEC-Backed Substrate

Next, the printed dipole is placed on dogbone-based HIS, as shown in Fig. 27. The dimension of the dipole is same as in Sect. 3.2.1. The dogbone dimensions are same as in Fig. 22b. The dipole is printed on a substrate (ε_r = 4.9, tan δ_e = 0.025, d = 0.0164λ). The lower substrate on which the dogbones are placed has ε_r = 4.9, tan δ_e = 0.025, d = 0.03λ. Figure 28 shows the resultant return loss less than −35 dB. Table 2 presents the return loss of same printed dipole placed on different combination of substrates. The gain of printed dipole placed on dogbone-based HIS is shown in Fig. 29. The antenna substrate is dielectric (ε_r = 4.9, tan δ_e = 0.025), while the substrate for dogbones is dielectric with ε_r = 6.16 and tan δ_e = 0.0025. The other dimensions are same as above.

3.2.4 HIS: Double-Layered Dogbones on PEC-Backed Substrate

Next, the dipole is printed on double layer of dogbone-based HIS, as shown in Fig. 30. The design parameters are kept same as in Sect. 3.2.1. The dogbones in the two layers are separated by 0.03λ. The thickness of dipole and dogbones is 0.35 mm.

The dipole substrate (ε_r = 4.9, tan δ_e = 0.025) has a thickness of 0.015λ. The lower substrate is taken as Rogers TM4 with ε_r = 4.5, tan δ_e = 0.002. Three configurations are considered.

(a) Dipole on double-layer PEC-backed HIS consisting of parallel dogbones (Fig. 30a),
(b) PEC-backed HIS with upper dogbone layer along the dipole length, whereas the lower dogbone layer perpendicular to it (Fig. 30b),
(c) PEC-backed HIS with upper dogbone layer perpendicular to the dipole length, whereas the lower dogbone layer along the dipole length (Fig. 30c).

The corresponding return loss of each configuration is shown in Fig. 31 respectively. It is apparent that the performance of printed dipole is good when HIS consists of parallel dogbones in upper and lower layers of the substrate. The gain of

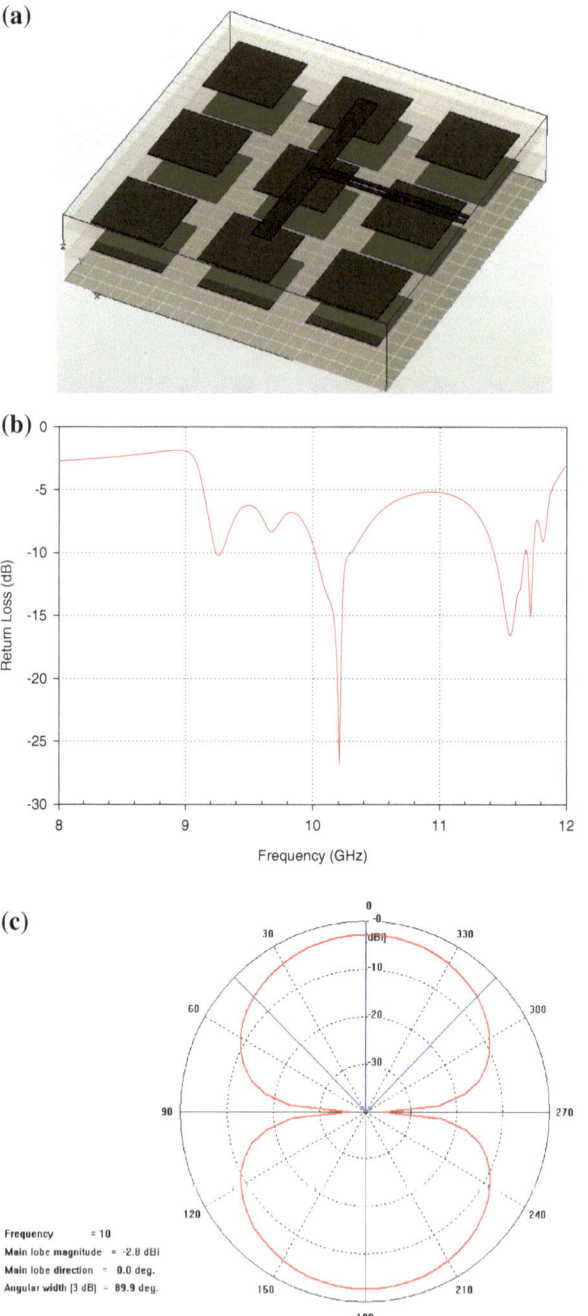

Fig. 26 **a** Printed dipole on a double-layered *square-patch* HIS. **b** Return loss of a printed dipole on a double-layered *square-patch* HIS. **c** Radiation pattern of a printed dipole on a double-layered *square-patch*-based HIS

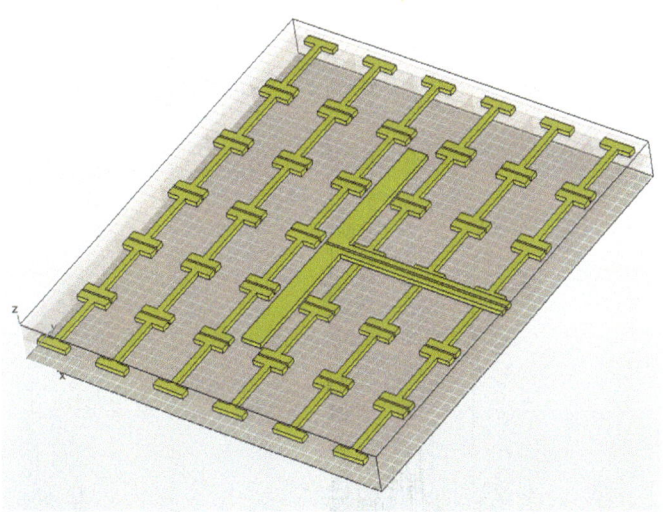

Fig. 27 A printed dipole antenna on HIS consisting of copper dogbones

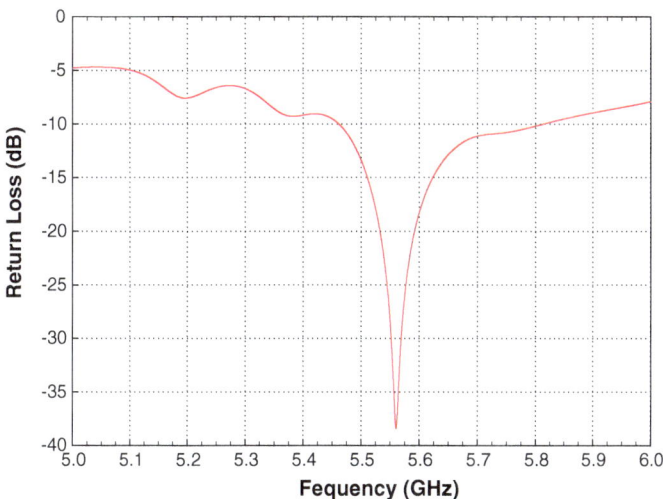

Fig. 28 Return loss of a printed dipole antenna on HIS consisting of copper dogbones

Table 2 Return loss for different combination of antenna substrate and HIS

ε_r of upper substrate	ε_r of lower substrate	Resonant frequency (GHz)	Return loss (dB)
4.9	4.9	5.75	53
2.2	4.9	5.86	43
6.15	4.9	5.41	37
4.9	6.15	5.28	28
4.5	4.9	5.5	37
4.82	4.9	5.56	22

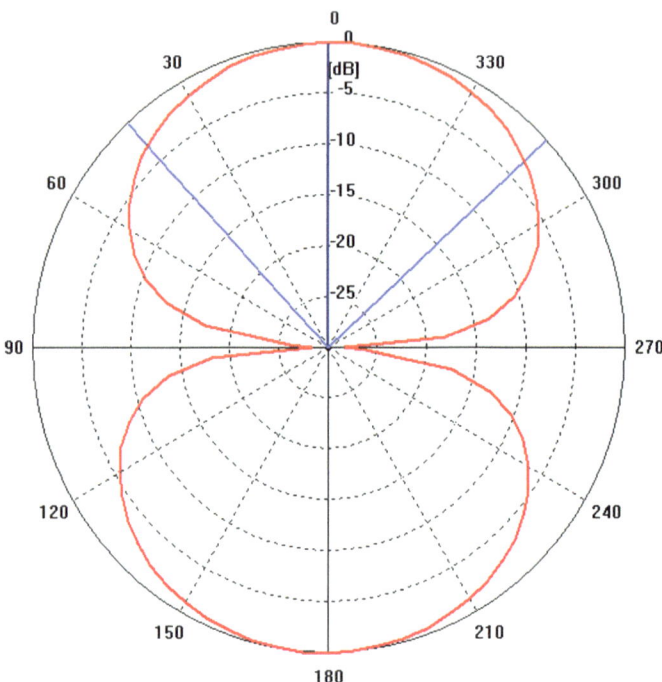

Fig. 29 Gain of a printed dipole antenna on HIS consisting of copper dogbones

the dipole antenna with cross double-layered dogbone-based HIS (Fig. 30c) is shown in Fig. 32.

The same design (Fig. 30c) is simulated in THz frequency range with the dimensions in μm. The corresponding return loss of the printed dipole is shown in Fig. 33.

The performance of the antenna can be further improved by appropriate choice of constitutive parameters of the substrates. The constitutive parameters of the substrate material are frequency-dependent. However, here the permittivity and loss tangent of the substrates at GHz are considered even for THz. This may be the reason for degradation in the return loss.

3.3 Folded Dipole

A folded dipole is known to have a higher input impedance and bandwidth as compared to a conventional printed dipole (Kraus et al. 2006). In this section, folded dipole is placed over single- and double-layered dogbone-based HIS. The model is simulated at 5.5 GHz frequency. The overall length of folded dipole is taken as 0.45λ, with 1 mm gap between the arms and 1 mm strip width (Fig. 34). The antenna is excited by a waveguide port through twin-line.

Fig. 30 A printed dipole on
double-layered copper
dogbone-based HIS.
a Parallel dogbones.
b Crossed dogbones; Dipole
aligned along the upper
dogbone layer. **c** Crossed
dogbones; Dipole aligned
along the lower dogbone layer

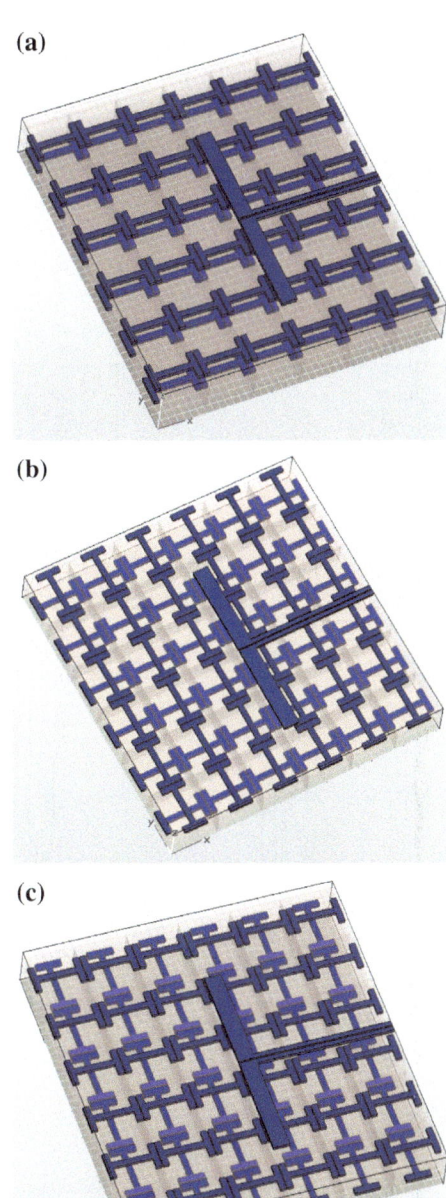

Fig. 31 Return loss of
double-layered
dogbone-based HIS
consisting of **a** Parallel
dogbones. **b** Crossed
dogbones; dipole aligned
along the *upper* dogbone
layer. **c** Crossed dogbones;
dipole aligned along the *lower*
dogbone layer

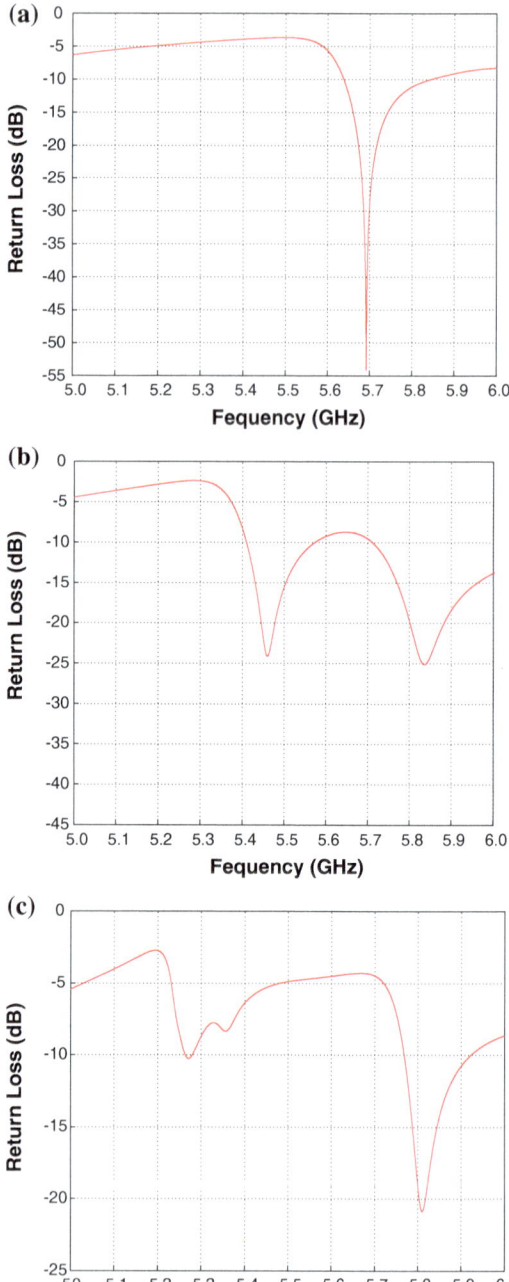

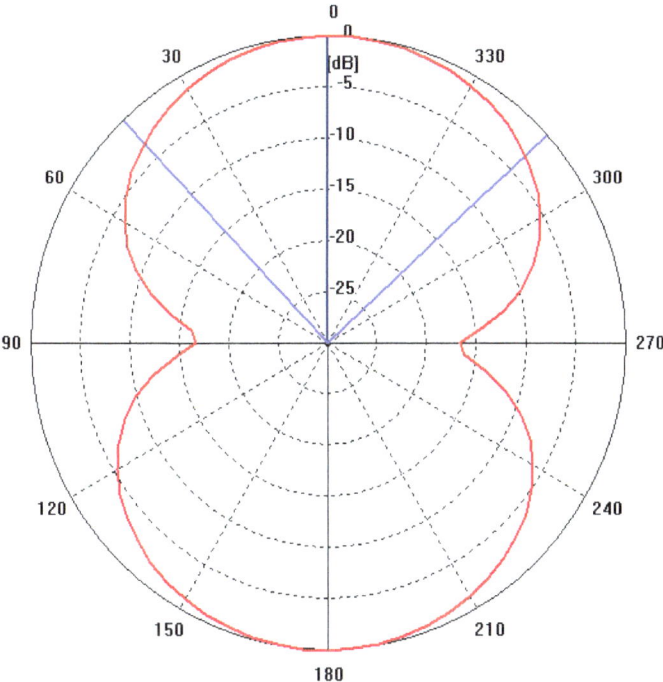

Fig. 32 Gain of printed dipole antenna on cross double dogbone layer HIS (Fig. 30c)

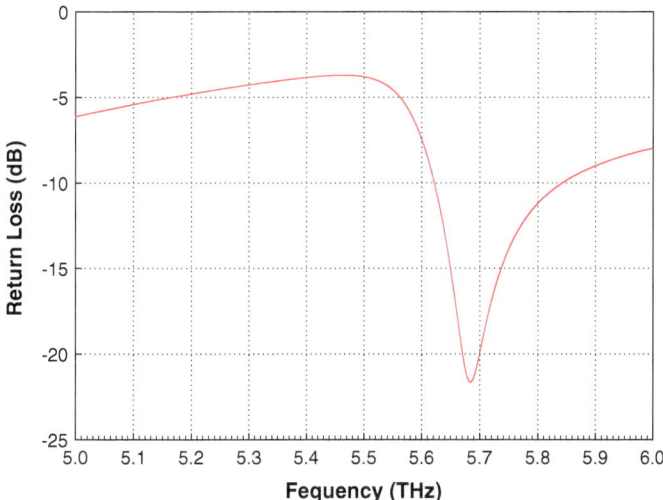

Fig. 33 Return loss of printed dipole antenna on double layer of dogbone-based HIS (Fig. 30c) in THZ region

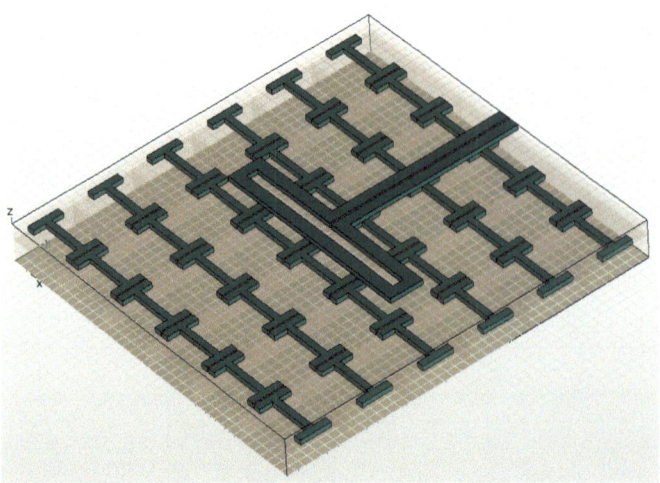

Fig. 34 A folded dipole antenna on HIS consisting of copper dogbones

3.3.1 Single-Layered HIS: Metallic Dogbones on PEC-Backed Substrate

The folded dipole is placed on a single-layered dogbone-based HIS, as shown in Fig. 34. The HIS consists of a dielectric substrate ($\varepsilon_r = 6.15$, $\tan \delta_e = 0.0025$) and an array of copper dogbones. The geometric parameters of dogbones are; $A = 0.128\lambda$, $B = 0.128\lambda$, $A1 = 0.016\lambda$, $A2 = 0.125\lambda$, $B1 = 0.064\lambda$, $B2 = 0.0128\lambda$, and $T = 0.35$ mm. The material of dipole is taken as copper. The dipole substrate is chosen as Rogers RT5870 ($\varepsilon_r = 2.33$, $\tan \delta_e = 0.0012$) with same thickness (T). The computed return loss for this design is shown in Fig. 35a. If the substrate of dipole is changed, i.e., the permittivity value is changed from 2.33 to 2.2, the return loss, and the resonant frequency would also change, as shown in Fig. 35b. Lower is the permittivity of the substrate, higher will be the resonant frequency of the dipole. The gain pattern of the folded dipole (Fig. 35b) is shown in Fig. 36.

3.3.2 Double-Layered HIS: Metallic Dogbones on PEC-Backed Substrate

As discussed above, a folded dipole is preferred over the conventional printed dipole for improved impedance matching. In this sub-section, the folded dipole is placed over double-layered dogbone-based HIS. There can be either parallel dogbones or crossed (perpendicular) dogbones in upper and lower layers of the substrate. Figure 37 shows the folded dipole on double-layered cross dogbone-based HIS. The geometrical parameters of dogbones, dipole, and substrate are same as in

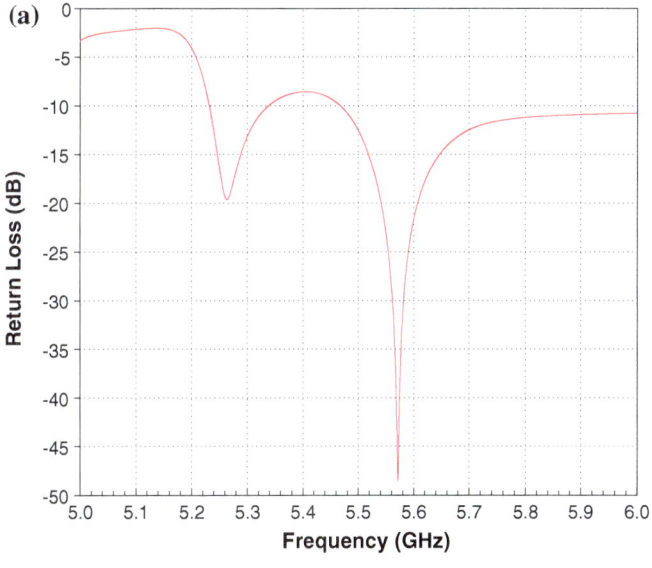

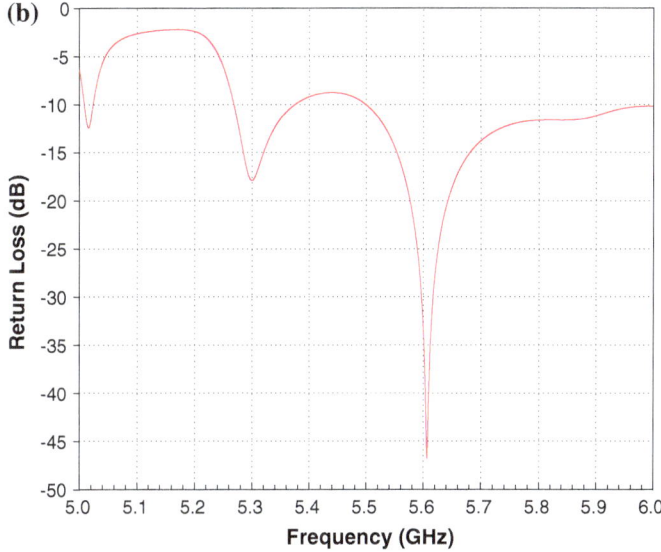

Fig. 35 Return loss of folded dipole antenna on HIS consisting of copper dogbones. **a** Dipole substrate's $\varepsilon_r = 2.33$, $\tan \delta_e = 0.0012$. **b** Dipole substrate's $\varepsilon_r = 2.2$, $\tan \delta_e = 0.0009$

previous section. The distance between two dogbones layer is 0.3λ. The constitutive parameters of the dielectric substrate on which the dogbones are placed are $\varepsilon_r = 6.15$ and $\tan \delta_e = 0.0025$.

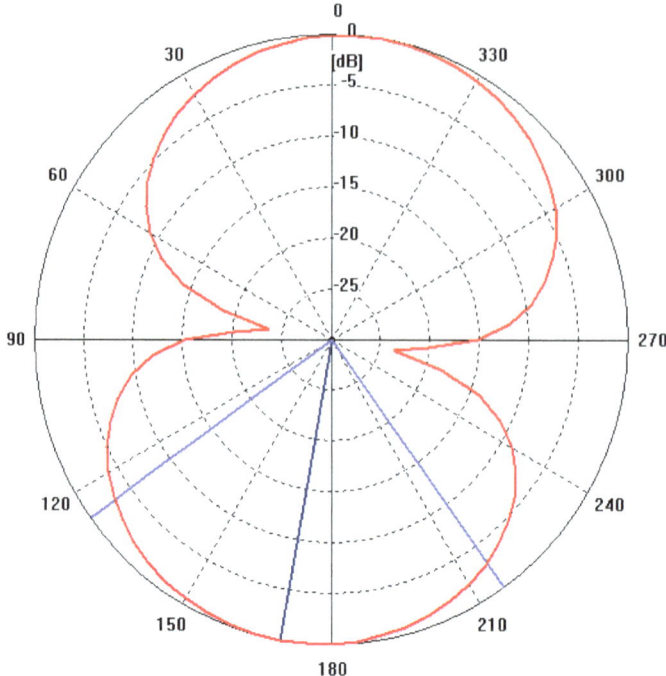

Fig. 36 Gain of folded dipole antenna on HIS (Fig. 35b)

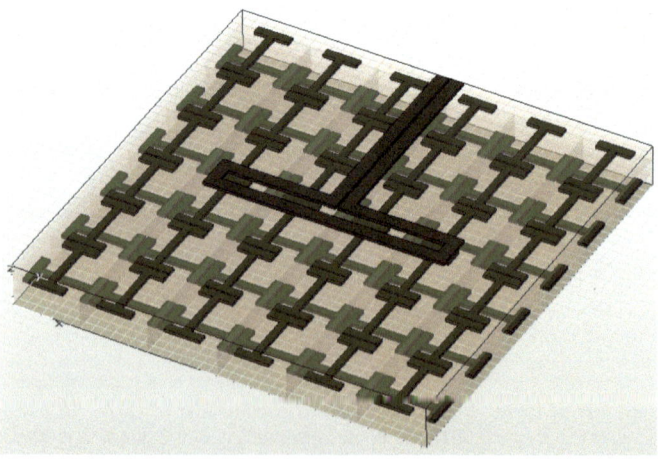

Fig. 37 Folded dipole antenna on cross double dogbone layer HIS

The lower dogbone layer is along the dipole, whereas the upper dogbone layer is perpendicular to it. The effect of dipole substrate on the performance of the design is analyzed by taking two different dipole substrates, keeping other parameters same.

Figure 38 presents the return loss of the folded dipole placed on two different substrates, one with $\varepsilon_r = 3$; $\tan \delta_e = 0.0013$ and other with $\varepsilon_r = 4.5$; $\tan \delta_e = 0.002$. It may be observed that higher permittivity shifts the resonant frequency toward lower frequency. Moreover, the return loss of the design got degraded.

Next, the dipole substrate ($\varepsilon_r = 3$; $\tan \delta_e = 0.0013$) is kept constant and HIS substrate is changed (Fig. 39). It may be seen that with increase in the permittivity of the lower substrate (i.e., from $\varepsilon_r = 6.15$; $\tan \delta_e = 0.0025$ to $\varepsilon_r = 9.2$; $\tan \delta_e = 0.0023$), the resonant frequency of the design got shifted toward lower frequency. However, the change in dogbone substrate does not affect the return loss significantly. The return loss remains at -50 dB or even better. Thus, it may be inferred that role of dipole substrate is more prominent in controlling return loss than the dogbone substrate. However, the resonant frequency of the design depends on both the substrates. The radiation pattern of folded dipole over double-layer cross dogbone-based HIS (Fig. 38b) is shown in Fig. 40.

Next, the folded dipole is placed on a double-layer parallel dogbone-based HIS (Fig. 41). The dimensions are same as shown in the Fig. 37. However, the orientation of dogbones is perpendicular to the dipole.

Figure 42 presents the return loss of the folded dipole for two different dipole substrates ($\varepsilon_r = 3$; $\tan \delta_e = 0.0013$ and $\varepsilon_r = 4.5$; $\tan \delta_e = 0.002$). Similar trend is observed as in the case of folded dipole on double-layered cross dogbone-based HIS (Fig. 38). If the performances of folded dipole is compared for parallel and cross double dogbone layer HIS, the cross dogbone configuration proves to be the better choice. However, this is opposite to the observation made in case of printed dipole design.

4   Low Profile Dipole Antenna on Non-planar High Impedance Substrate

In this section, non-planar HIS is designed using dogbones. A cylindrical dielectric surface is taken over which the metallic dogbones are printed. Then, the dipole along with its dielectric substrate is printed over the curved HIS. The dipole is made conformal to the surface.

4.1   Printed Dipole on a Single-Layered Dogbone-Based HIS

The first step is to design non-planar HIS. A cylindrical section of dielectric ($\varepsilon_r = 10.2$, $\tan \delta = 0.003$) with radius, $r = 7.7986$ mm; $h = 24.5$ mm; $t = 1.61$ mm is chosen as a substrate. A 3×3 dogbone array is printed on over the substrate, as shown in Fig. 43. The dimensions of dogbones are same as in previous sections. The gap between the dogbones in radial direction is 0.88 mm (approx.). The rows

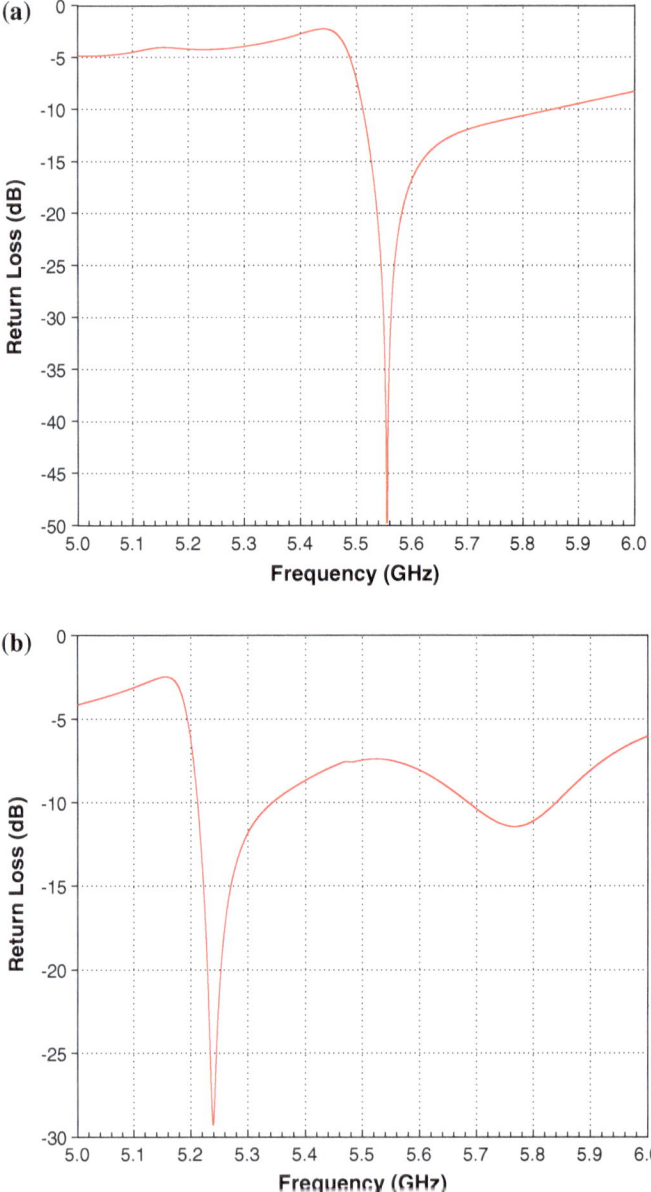

Fig. 38 Return loss of a folded dipole antenna on cross double dogbone layer HIS. **a** Dipole substrate's $\varepsilon_r = 3$, tan $\delta_e = 0.0013$. **b** Dipole substrate's $\varepsilon_r = 4.5$, tan $\delta_e = 0.002$

of dogbones are separated by distance of 7.18 mm. The HIS is backed by a rectangular PEC ground plane. The design is excited using waveguide port placed at $z = 11.08$ mm. Figure 44 shows the phase of the reflection coefficient of the design.

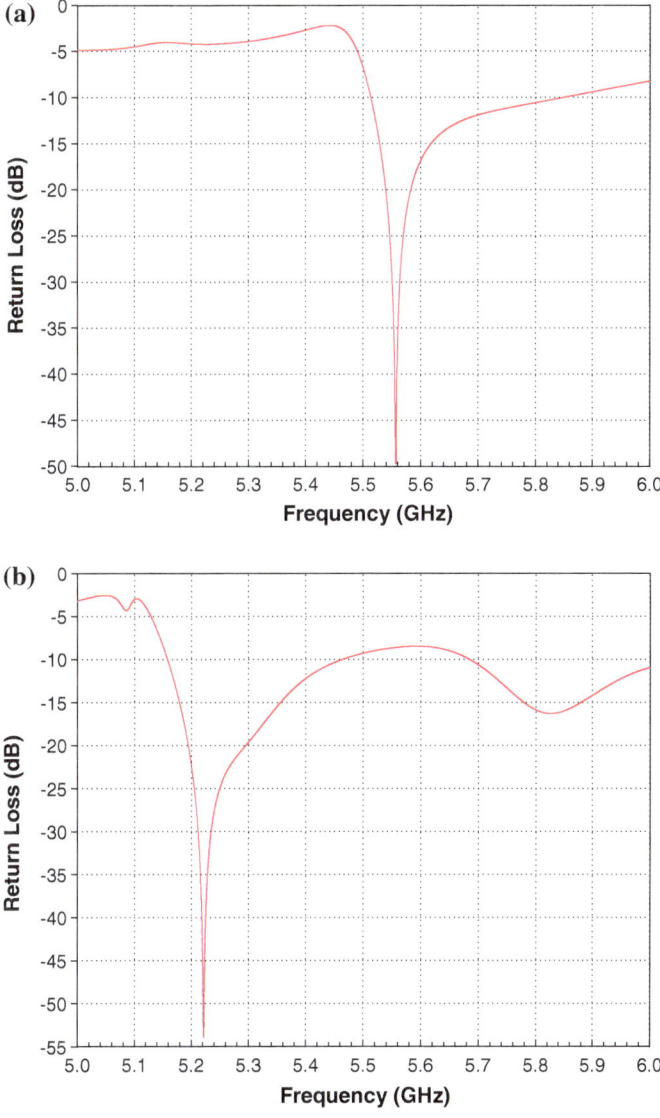

Fig. 39 Return loss of a folded dipole antenna on cross double dogbone layer HIS with dipole substrate (ε_r = 3, tan δ_e = 0.0013). **a** Dogbone substrate's ε_r = 6.15, tan δ_e = 0.0025. **b** Dogbone substrate's ε_r = 9.2, tan δ_e = 0.0023

Next step is to place dipole over a single-layered dogbone-based non-planar HIS (Fig. 45). Both the dogbone and the dipole substrate are taken as dielectric (ε_r = 4.9, tan δ = 0.025). The dipole excitation is given through a twin-line using waveguide port. The design frequency is 10 GHz.

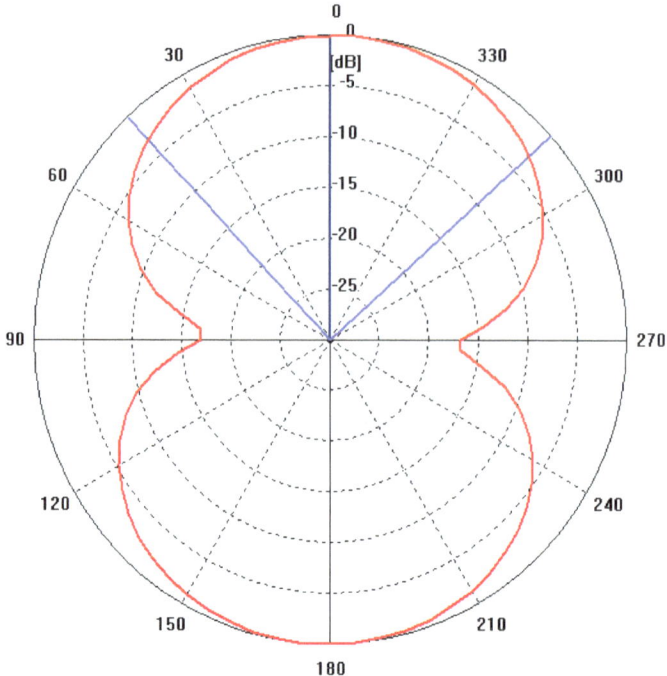

Fig. 40 Gain of folded dipole antenna on double-layered cross dogbone-based HIS (Fig. 38b)

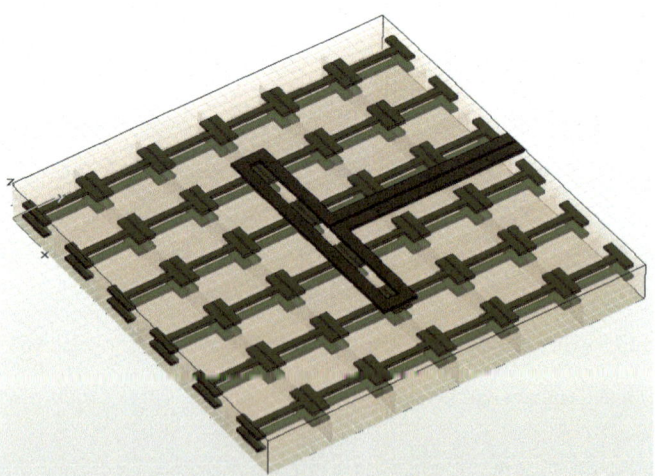

Fig. 41 Folded antenna on HIS consisting of double layer of parallel dogbones

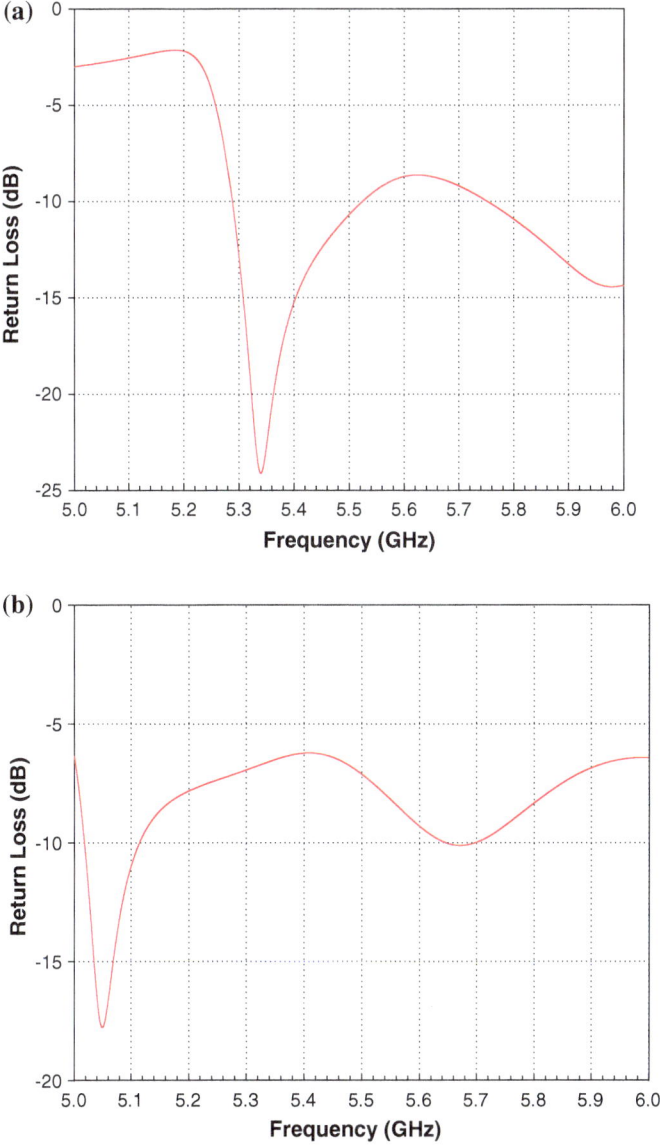

Fig. 42 Return loss folded dipole antenna on HIS consisting of copper dogbones. **a** Dipole substrate's $\varepsilon_r = 3$, $\tan \delta_e = 0.0013$. **b** Dipole substrate's $\varepsilon_r = 4.5$, $\tan \delta_e = 0.002$

The return loss of -22 dB is obtained at 8.75 GHz, as shown in Fig. 46. The corresponding radiation pattern for the design is shown in Fig. 47. The mainlobe gain of the dipole is obtained as -10 dBi.

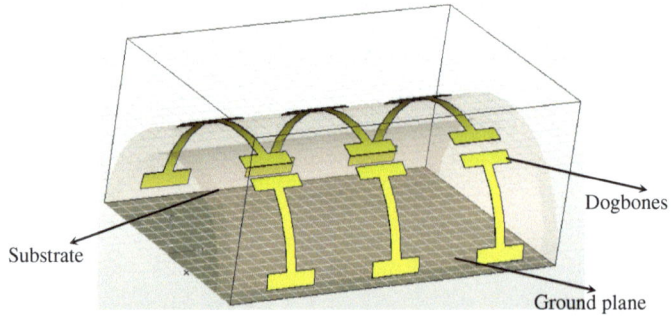

Fig. 43 Single-layered dogbone-based non-planar HIS

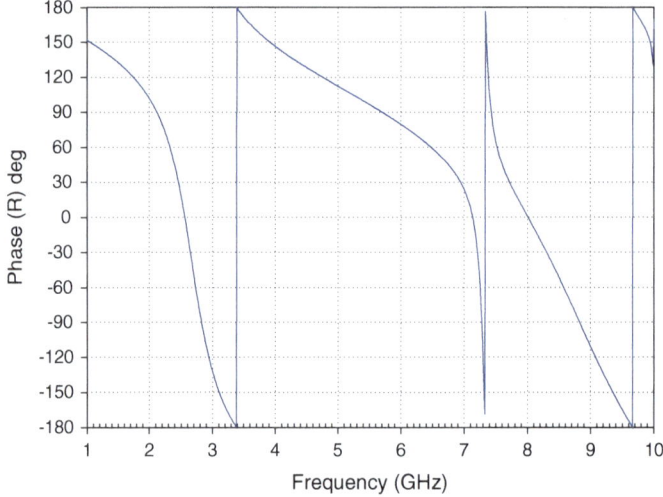

Fig. 44 Phase of reflection coefficient at the top of single-layered dogbone-based non-planar HIS

4.2 Folded Dipole on Double-Layered Dogbone-Based HIS

The single-layered dogbone-based non-planar HIS is modified into double-layered HIS (Fig. 48a). The dogbone substrate is a dielectric (ε_r = 10.2, tan δ = 0.003). The dimensions of the design are kept same as in Sect. 4.1. The radii of the outer and inner substrate layers are $r + t$ and r respectively, as shown in Fig. 48b. A rectangular PEC ground plane is placed below the HIS. Figure 49 shows the reflection coefficient (phase) of the design.

A folded dipole is placed over the double-layered dogbone-based non-planar HIS (Fig. 50).

The HIS substrate is a dielectric material with ε_r = 4.9, tan δ = 0.025, t = 1.61 mm. The dipole substrate is a dielectric material with ε_r = 2.2,

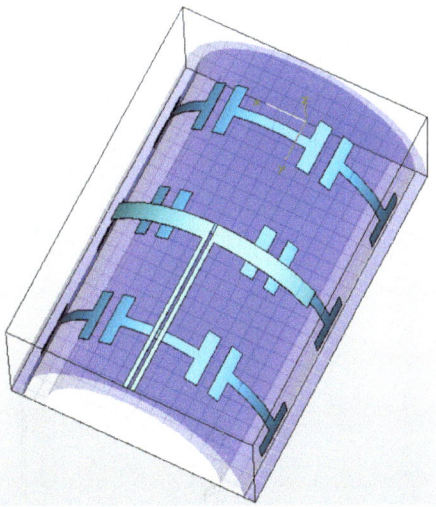

Fig. 45 Printed dipole on a single-layered non-planar dogbone HIS

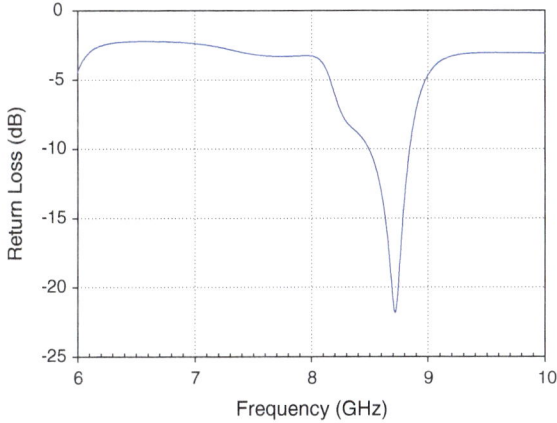

Fig. 46 Return loss of a printed dipole on a single-layered dogbone non-planar HIS

$\tan \delta = 0.0009$, $t = 0.035$ mm. The design frequency is 10 GHz. The dimensions of folded dipole are $L = 0.46\lambda$, $W = 0.033\lambda$, thickness = 0.035 mm. The antenna is excited using waveguide port. The return loss of the design is obtained as -22 dB (Fig. 51). The corresponding radiation pattern is shown in Fig. 52. The mainlobe gain of the folded dipole is obtained as 4.1 dBi.

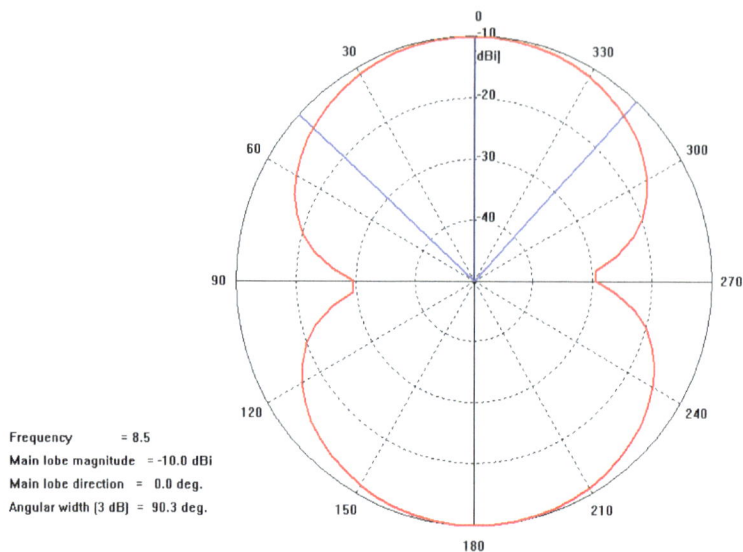

Fig. 47 Radiation pattern of a printed dipole on a single-layered dogbone-based non-planar HIS

5   Low Profile Dipole Array on Planar and Non-planar Dogbone-Based HIS

Next, the array of dipoles are designed on planar and non-planar HIS. The performance of the array design is analyzed on the basis of return loss and the radiation behavior.

5.1   Dogbone-Based Double-Layered Planar HIS

A double-layered dogbone-based HIS is considered for the substrate. The HIS substrate is dielectric with $\varepsilon_r = 10.2$, $\tan \delta = 0.0035$. The dipole substrate is another dielectric with $\varepsilon_r = 9.2$, $\tan \delta = 0.0023$. The dimensions of dogbones, substrate, and the printed dipole are kept same as in previous sections.

5.1.1   Two-Element Printed Dipole Array

First array of two printed dipoles is designed on double-layered dogbone-based HIS, shown in Fig. 53. The dipoles are excited using single waveguide port. The return loss of 2-element dipole array over planar dogbone-based HIS is obtained as −66 dB, as shown in Fig. 54. The radiation pattern of the array is shown in Fig. 55, with mainlobe gain of −5.5 dBi.

(a)

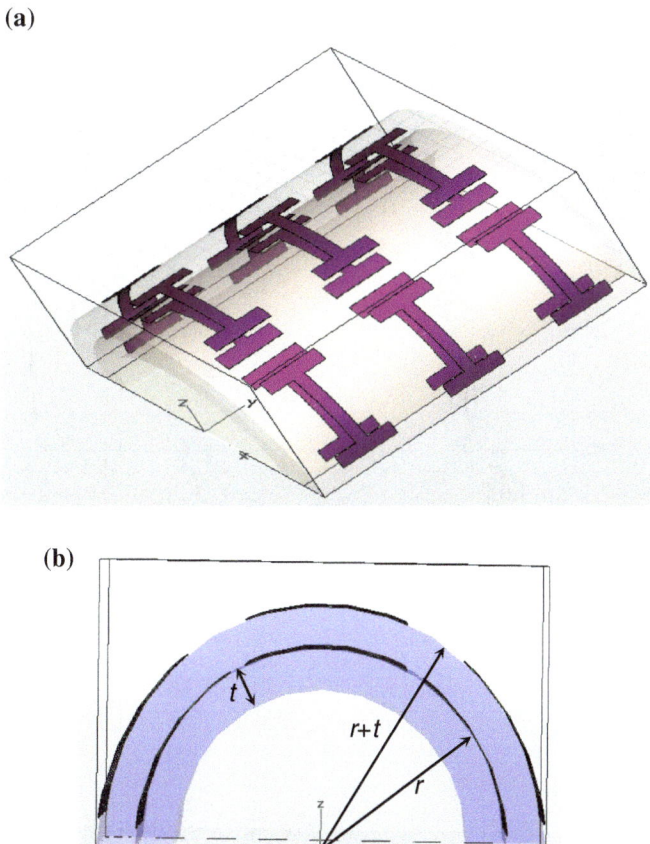

(b)

Fig. 48 Double-layered dogbone-based non-planar HIS. **a** Model. **b** Schematic

5.1.2 Three-Element Printed Dipole Array

Next, the number of elements is increased to three. The design of 3-element printed dipole array on double-layered dogbone-based planar HIS is shown in Fig. 56. All the design parameters are kept same as in 2-element dipole array design.

The design frequency is 10 GHz. The return loss of 3-element dipole array over double-layered dogbone-based planar HIS is obtained as −68 dB, shown in Fig. 57. The radiation pattern of the array is shown in Fig. 58, with mainlobe gain of −5.2 dBi.

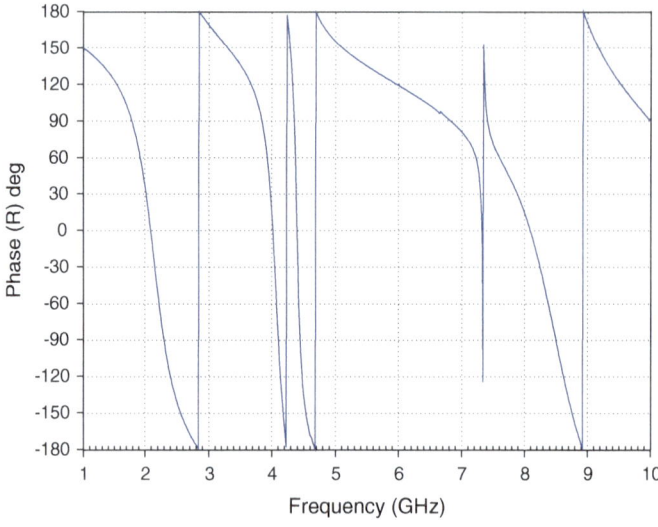

Fig. 49 Phase of reflection coefficient at the top of double-layered dogbone-based non-planar HIS

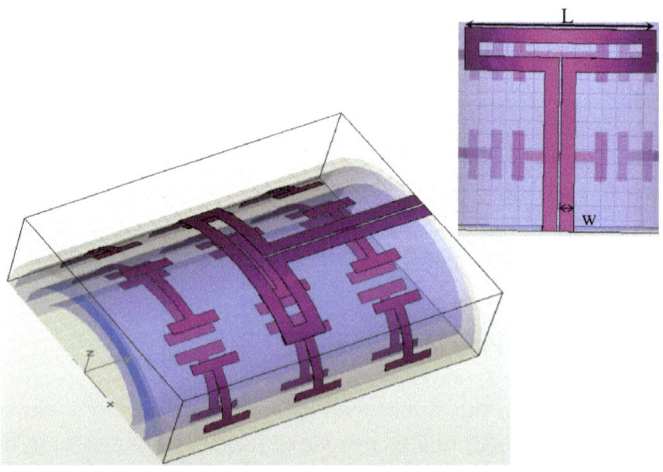

Fig. 50 Folded dipole on a double-layered dogbone-based non-planar HIS

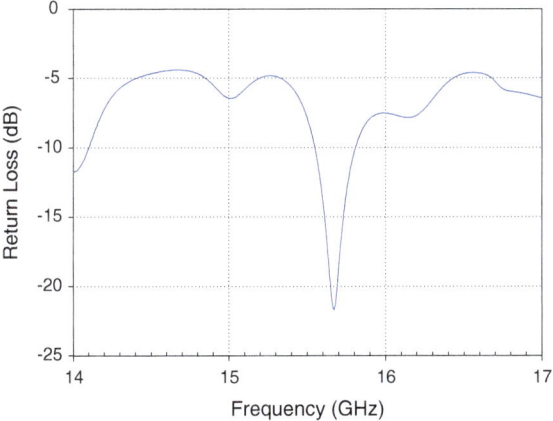

Fig. 51 Return loss of a folded dipole on a double-layered dogbone curved HIS

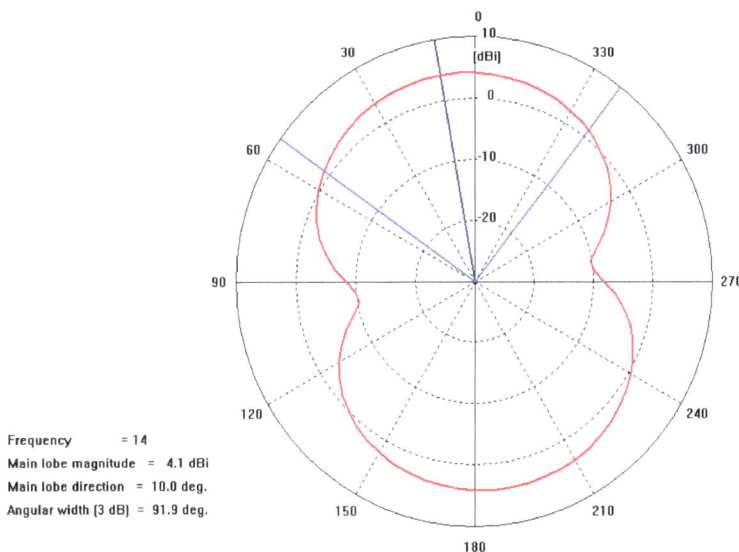

Fig. 52 Radiation pattern of a folded dipole on a double-layered dogbone-based non-planar HIS

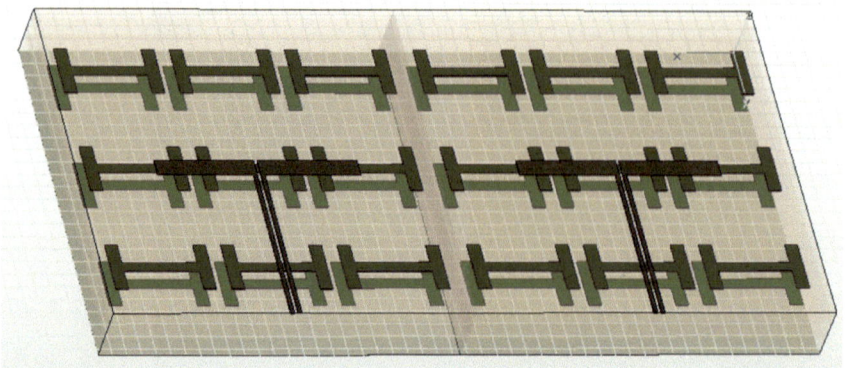

Fig. 53 2-Element printed dipole array on a double-layered dogbone-based planar HIS

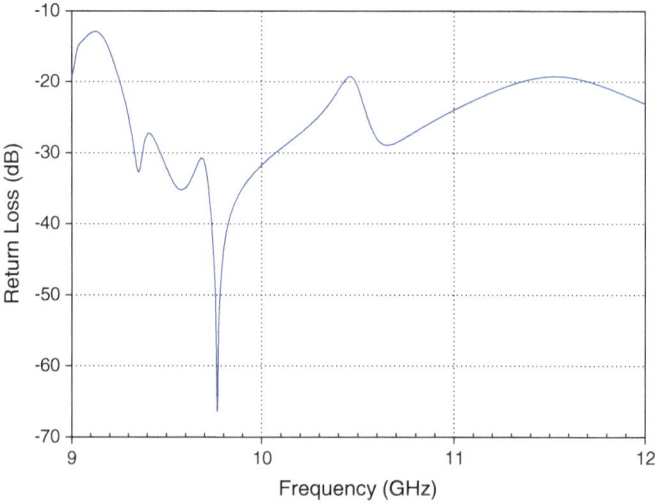

Fig. 54 Return loss of a 2-element printed dipole array on a double-layered dogbone-based planar HIS

5.2 Dogbone-Based Double-Layered Non-planar HIS

Once design of dipole array over planar HIS is done, the design is extended to non-planar HIS, discussed in Sect. 4.2.

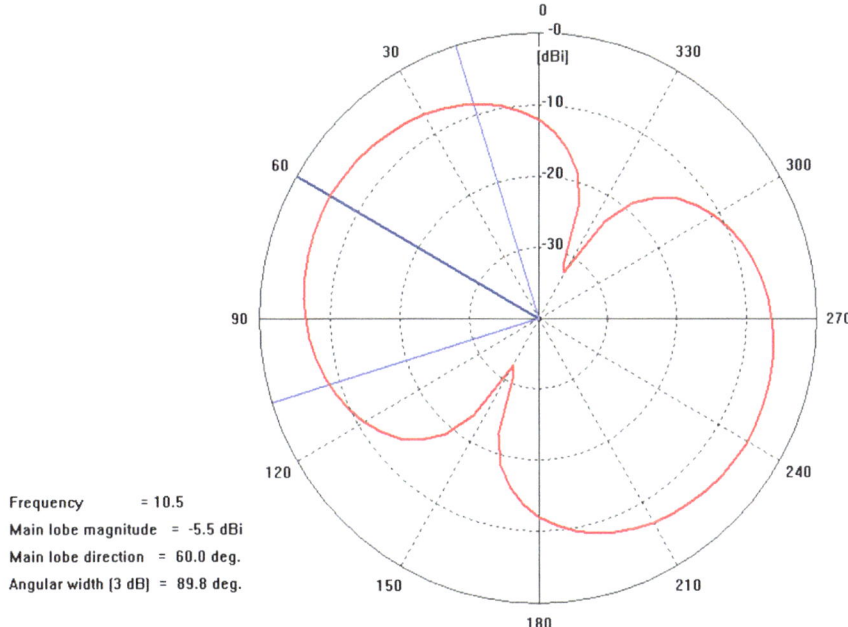

Frequency   = 10.5
Main lobe magnitude = -5.5 dBi
Main lobe direction = 60.0 deg.
Angular width (3 dB) = 89.8 deg.

Fig. 55 Radiation pattern of a 2-element printed dipole array on a double-layered dogbone-based planar HIS

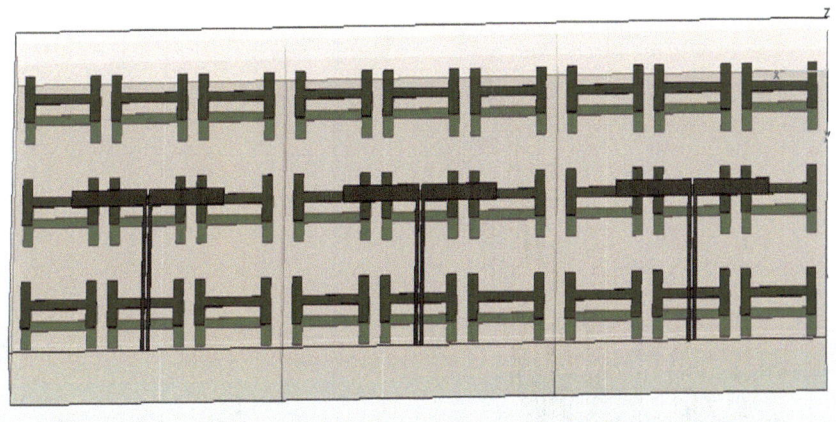

Fig. 56 3-Element printed dipole array on a double-layered dogbone-based planar HIS

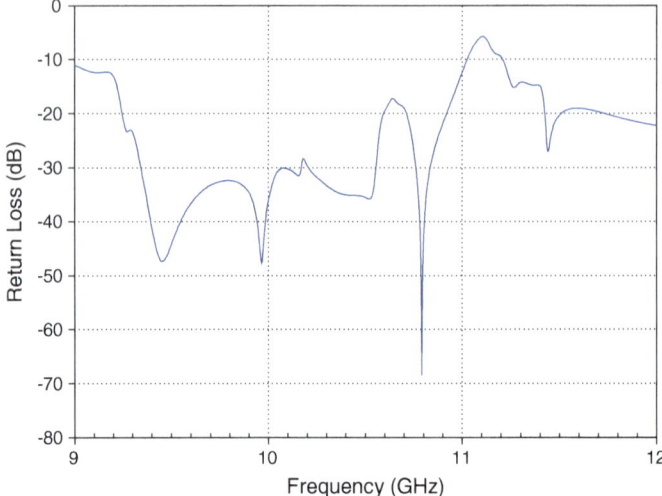

Fig. 57 Return loss of a 3-element printed dipole array on a double-layered dogbone-based planar HIS

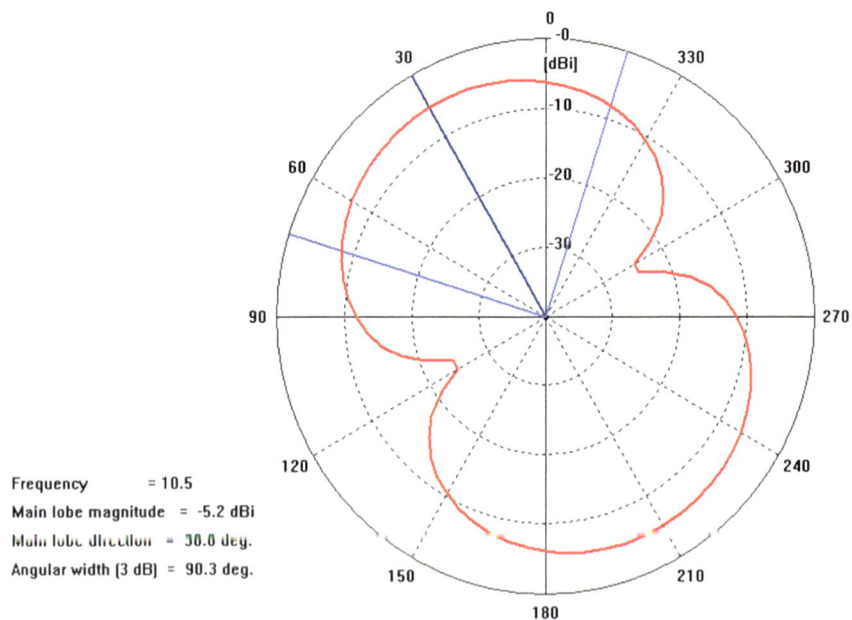

Fig. 58 Radiation pattern of a 3-element printed dipole array on a double-layered dogbone-based planar HIS

5.2.1   Two-Element Folded Dipole Array

Here, array of two-folded dipoles is designed on a double-layered dogbone-based non-planar HIS, shown in Fig. 59. The HIS substrate and dipole substrate are dielectric having $\varepsilon_r = 10.2$, $\tan\delta = 0.003$ and $\varepsilon_r = 4.9$, $\tan\delta = 0.025$ respectively. The radius of the cylindrical section of substrate is 15.59 mm.

A single waveguide port connecting the end of both the folded dipoles is used for excitation. The dimensions of dipole and substrate are kept same as in Sect. 4.2. The return loss of 2-element folded dipole array over double-layered dogbone-based non-planar HIS is obtained as −32 dB, as shown in Fig. 60. The radiation pattern of the array is shown in Fig. 61, with mainlobe gain of −4.6 dBi.

5.2.2   Three-Element Folded Dipole Array

Next, 3-element folded dipole array is designed on double-layered curved HIS (Fig. 62). The design frequency is 10 GHz. The HIS substrate and dipole substrate are taken same as in above section. Single waveguide port excitation is given to three dipoles.

The return loss of 3-element folded dipole array over double- layered non-planar HIS is obtained as −33 dB, shown in Fig. 63. The radiation pattern obtained is shown in Fig. 64, with mainlobe gain of −2.1 dBi.

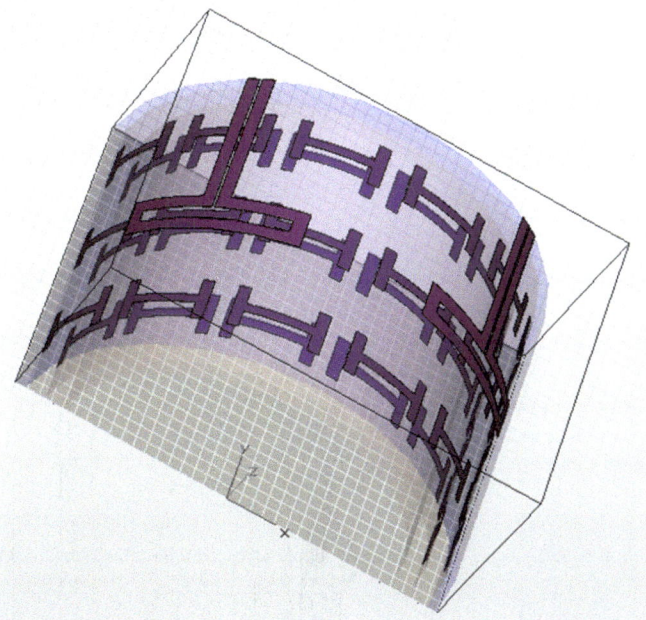

Fig. 59   2-Element folded dipole array on a double-layered dogbone-based non-planar HIS

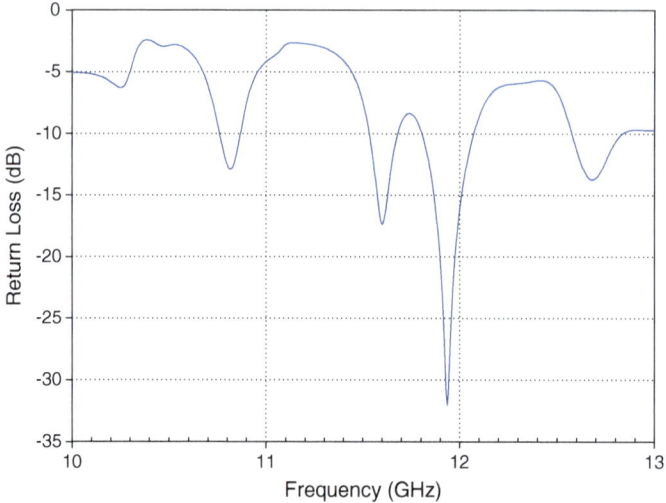

Fig. 60 Return loss of a 2-element folded dipole array on a double-layered dogbone-based non-planar HIS

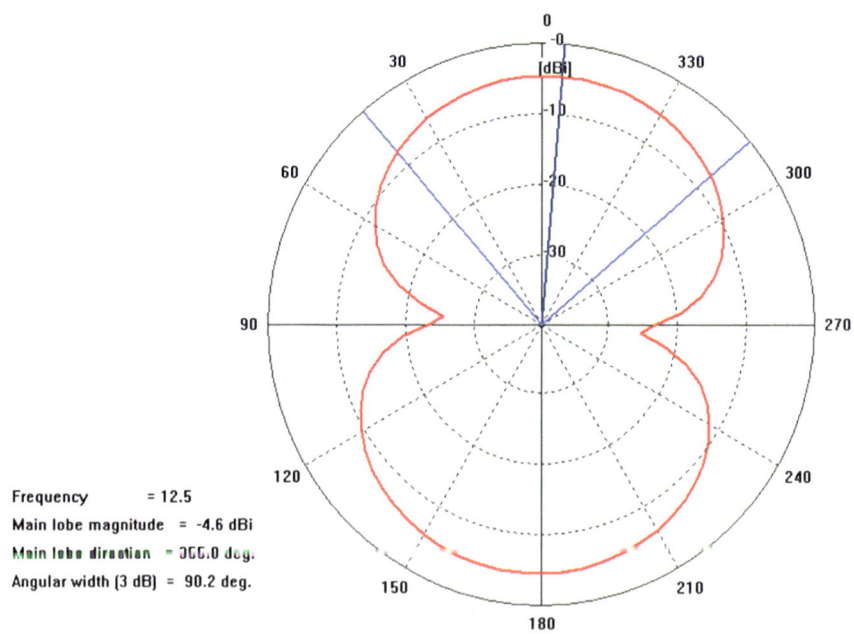

Fig. 61 Radiation pattern of a 2-element folded dipole array on a double-layered dogbone-based non-planar HIS

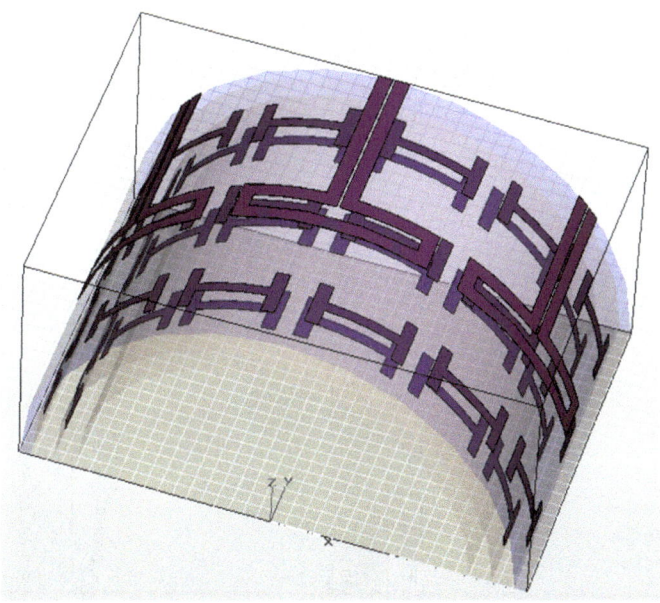

Fig. 62 3-Element folded dipole array on a double-layered dogbone-based non-planar HIS

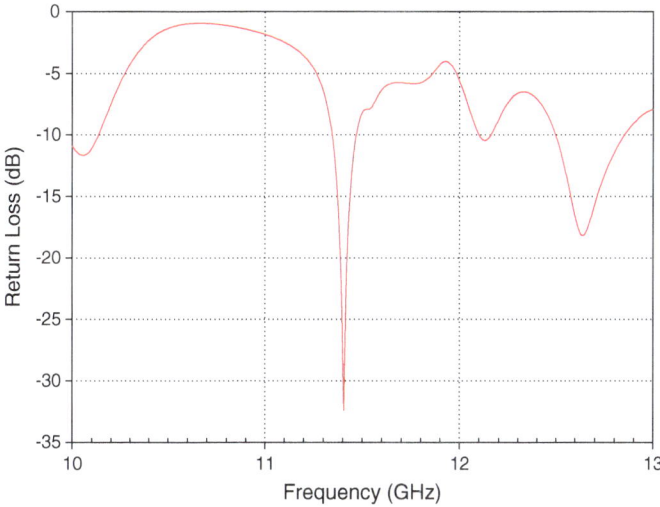

Fig. 63 Return loss of a 3-element folded dipole array on a double-layered dogbone-based non-planar HIS

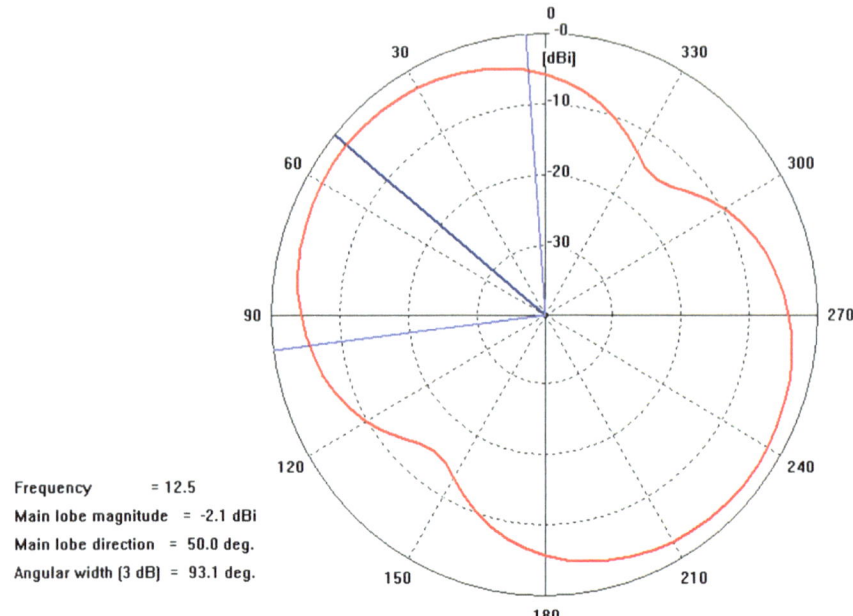

Frequency   = 12.5
Main lobe magnitude = -2.1 dBi
Main lobe direction = 50.0 deg.
Angular width (3 dB) = 93.1 deg.

Fig. 64 Radiation pattern of a 3-element folded dipole array on a double-layered dogbone-based non-planar HIS

The design presented shows acceptable return loss and mainlobe gain of radiation pattern. The design study carried out is geared toward the EM design and analysis of conformal array.

6 Conclusion

The high impedance substrate is a preferred substrate for low profile antenna design. Its unique boundary conditions make image currents in-phase. This permits radiating elements to be printed on them, without any disturbance in their radiation. It provides better impedance matching, enhanced bandwidth, broadside directivity owing to total reflection from the reactive surface, and high input impedance.

If the dipole is placed over HIS, the resonant frequency of dipole is lowered. This is due to the compensation of dipole's inductive behavior with the capacitive one. This reduces the antenna dimensions in terms of wavelength, and hence facilitates the antenna miniaturization.

The performance of HIS is evaluated in terms of parameters like antenna gain, magnitude and phase of the reflection coefficient of the design. The results for two types of HIS, square-patch-based HIS and dogbone-based HIS are discussed. It is observed that it is not only improved return loss, but also wider bandwidth is achieved with dogbone-based HIS, as compared to square patch-based HIS.

The role of dipole and dogbone substrate is analyzed in deciding the antenna performance. Both single-layered and double-layered dogbone-based HIS is considered. If the substrate of dipole is changed, the return loss and the resonant frequency also changes. Lower is the permittivity of the substrate, higher will be the resonant frequency of the dipole. However, this is not exactly true for dogbone substrate. The resonant frequency of dipole shifts with change in dogbone substrate, but there is no significant variation in the return loss. In other words, the role of dipole substrate is more prominent in controlling return loss than the dogbone substrate. The resonant frequency of the dipole depends on both the substrates.

The performance of printed dipole is observed to be better when parallel dogbone-based double layer HIS is considered. This is however not true for folded dipole design. The return loss of folded dipole is better in case of cross double dogbone layer HIS. Thus, one may infer that the choice of configuration of HIS depends on the type of radiating element as well. The study carried out is geared toward the design of dipole array over HIS. The HIS can be designed using other shapes of metallic conductors.

References

Azad, M.Z., and M. Ali. 2008. Novel wideband directional dipole antenna on a mushroom like EBG structure. *IEEE Transactions on Antennas and Propagation* 56(5): 1242–1250.

Best, S.R., and D.L. Hanna. 2008. Design of broadband dipole in close proximity to an EBG ground plane. *IEEE Antennas and Propagation Magazine* 50(6): 52–64.

Colburn, J.S., D.F. Sievenpiper, B.H. Fong, J.J. Ottusch, J.L. Visher, and P.R. Herz. 2007. Advances in artificial impedance surface conformal antennas. IEEE Antennas and Propagation Society International Symposium, Honolulu, Hawaii, USA, pp. 3820–3823.

Cure, D., T.M. Weller, and F.A. Miranda. 2013. Study of a low-profile 2.4 GHz planar dipole antenna using a high-impedance surface with 1-D varactor tuning. *IEEE Transactions on Antennas and Propagation* 61(2): 506–515.

Donzelli, G., A. Vallecchi, F. Capolino, and A. Schuchinsky. 2009. Anisotropic metamaterials made of paired conductors: Particle resonances, phenomena and properties. *Metamaterials* 3 (1): 10–27.

Erturk, V.B., and B. Guner. 2004. Finite phased arrays of printed dipoles on large circular cylinders: A comparison with the planar case. Proceedings of URSI/International Symposium of Electromagnetic Theory, Pisa, Italy, pp. 972–974.

Kraus, J.D., R.J. Marhefka, and A.S. Khan. 2006. *Antennas for all applications,* 3rd edn, 962 p. New Delhi: Tata McGraw-Hill. ISBN-13:978-0-07-060185-7.

McVay, J., A. Hoorfar, and N. Engheta. 2003. Radiation characteristics of microstrip dipole antennas over a high-impedance metamaterial surface made of Hilbert Inclusions. IEEE MTT-S International Microwave Symposium Digest, Philadelphia, PA, USA, vol. 1, pp. 587–590.

Mosallaei, H., and K. Sarabandi. 2004. Antenna miniaturization and bandwidth enhancement using a reactive impedance substrate. *IEEE Transactions on Antennas and Propagation* 52(9): 2403–2414.

Sievenpiper, D., Z. Lijun, R.F.J. Broas, N.G. Alexopolous, and E. Yablonovitch. 1999. High-impedance electromagnetic surfaces with a forbidden frequency band. *IEEE Transactions on Microwave Theory and Techniques* 47: 2059–2074.

Vallecchi, A., and F. Capolino. 2009a. Metamaterials based on pairs of tightly-coupled scatterers. in *Theory and phenomena of metamaterials*, Chap. 19. Boca Raton: CRC Press. ISBN:978-1420054255.

Vallecchi, A., and F. Capolino. 2009b. Thin high-impedance metamaterials substrate and its use in low profile antennas suitable for system integration. Proceedings of European Conference on Antennas and Propagation (EUCAP), Berlin, Germany, pp. 861–865.

Vallecchi, A., J.R.D. Luis, F. Capolino, and F.D. Flaviis. 2012. Low profile fully planar folded dipole antenna on a high impedance surface. *IEEE Transactions on Antennas and Propagation* 60(1): 51–62.

About the Book

This book presents electromagnetic (EM) design and analysis of dipole antenna array over high impedance substrate (HIS). HIS is a preferred substrate for low-profile antenna design, owing to its unique boundary conditions. Such substrates permit radiating elements to be printed on them without any disturbance in the radiation characteristics. Moreover HIS provides improved impedance matching, enhanced bandwidth, and increased broadside directivity owing to total reflection from the reactive surface and high input impedance. This book considers different configurations of HIS for array design on planar and non-planar high-impedance surfaces. Results are presented for cylindrical dipole, printed dipole, and folded dipole over single- and double-layered square-patch-based HIS and dogbone-based HIS. The performance of antenna arrays is analyzed in terms of performance parameters such as return loss and radiation pattern. The design presented shows acceptable return loss and mainlobe gain of radiation pattern. This book provides an insight to EM design and analysis of conformal arrays. This book serves as an introduction for beginners in the design and analysis of HIS-based antenna arrays. It includes pictorial description of both planar and non-planar array design and the detailed discussion of the performance analysis of HIS-based planar and non-planar antenna array. It will prove useful to researchers and professionals, alike.

© The Author(s) 2016   51
H. Singh et al., *Low Profile Conformal Antenna Arrays*
on High Impedance Substrate, SpringerBriefs in Computational Electromagnetics,
DOI 10.1007/978-981-287-763-5

Author Index

A
Alexopolous, N.G., 3
Ali, M., 3
Azad, M.Z., 3

B
Best, S.R., 1
Broas, R.F.J., 3

C
Capolino, F., 1, 3
Colburn, J.S., 2
Cure, D., 2

D
Donzelli, G., 2, 3, 11

E
Engheta, N., 2
Erturk, V.B., 2

F
Flaviis, F.D., 2
Fong, B.H., 2

G
Guner, B., 2

H
Hanna, D.L., 1
Herz, P.R., 2
Hoorfar, A., 2

K
Khan, A.S., 24
Kraus, J.D., 15, 24

L
Lijun, Z., 3
Luis, J.R.D., 3, 11, 15

M
Marhefka, R.J., 24
McVay, J., 2
Miranda, F.A., 2
Mosallaei, H., 2

O
Ottusch, J.J., 1

S
Sarabandi, K., 2
Schuchinsky, A., 3, 11
Sievenpiper, D., 3
Sievenpiper, D.F., 2

V
Vallecchi, A., 1–3, 11, 15
Visher, J.L., 2

W
Weller, T.M., 2

Y
Yablonovitch, E., 3

© The Author(s) 2016
H. Singh et al., *Low Profile Conformal Antenna Arrays*
on High Impedance Substrate, SpringerBriefs in Computational Electromagnetics,
DOI 10.1007/978-981-287-763-5

Subject Index

A
Artificial magnetic conductor, 1, 11
Antenna efficiency, 3, 18

B
Bandwidth, 2, 11, 14, 18, 24, 48
Boundary conditions, 4, 48

C
Conformal antenna array, 2
Constitutive parameters, 11, 24, 29
Current distribution, 3
Cylindrical dipole, 2, 15, 17, 18

D
Design frequency, 10, 17, 18, 33, 37, 39, 45
Directivity, 2, 3, 15, 48
Dogbone, 2–4, 8, 11, 13, 17, 21, 24, 28, 31, 36, 39

E
Electromagnetic bandgap, 1

F
Folded dipole, 2, 3, 24–27, 36, 37, 45

G
Ground plane, 2, 8, 10, 32, 36

H
High impedance substrate, 1–3, 6, 8, 15, 18, 24, 28, 31, 36, 38, 35
  non-planar, 2, 31, 36, 38, 42
  planar, 6, 38, 44

I
Image current, 2, 3, 48
Impedance matching, 2–3, 11, 15, 18, 28, 48
Input impedance, 3, 11, 24, 48
Interference, 2
  constructive, 2, 3
  destructive, 2

L
Leaky waves, 3
Loss tangent, 8, 11, 15, 24
Low profile, 1–4, 48

M
Magnetic resonance, 3, 11, 15
Mainlobe, 20, 21, 35, 37, 38, 45
Metamaterial, 1–3
Microstrip patch antenna, 2

P
Perfect magnetic conductor, 3, 15
Printed dipole, 2, 18, 20, 21, 24, 31, 38

R
Radiating current, 2
Radiation efficiency, 3
Radiation pattern, 2, 3, 15, 20, 31, 35, 37–39, 45
Reflection coefficient, 2, 4, 6, 8, 11, 14, 32, 36, 48
Resonant frequency, 5, 8, 15, 28, 48
Return loss, 2, 4, 15, 18, 20, 21, 23, 28, 31, 35, 37, 45

© The Author(s) 2016
H. Singh et al., *Low Profile Conformal Antenna Arrays*
on High Impedance Substrate, SpringerBriefs in Computational Electromagnetics,
DOI 10.1007/978-981-287-763-5

S
Square patch, 2, 6, 15, 18, 48
Substrate, 2, 4, 6, 8, 11, 14, 15, 17, 18, 21, 28,
    31, 38, 45
Surface reactance, 4
Surface waves, 3

U
Unit cell, 2, 4

W
Waveguide port, 4, 6, 8, 11, 20, 24, 32, 37, 38,
    45